UNLOCKING
Creativity

UNLOCKING
Creativity

HOW TO SOLVE ANY PROBLEM
AND MAKE THE BEST DECISIONS
BY SHIFTING CREATIVE MINDSETS

Michael A. Roberto

WILEY

Published by John Wiley & Sons, Inc., Hoboken, New Jersey.
Published simultaneously in Canada.

For general information on our other products and services or for technical support, please contact our
Customer Care Department within the United States at (800) 762–2974, outside the United States at
(317) 572–3993 or fax (317) 572–4002.

Wiley publishes in a variety of print and electronic formats and by print-on-demand. Some material
included with standard print versions of this book may not be included in e-books or in
print-on-demand. If this book refers to media such as a CD or DVD that is not included in the
version you purchased, you may download this material at http://booksupport.wiley.com. For more
information about Wiley products, visit www.wiley.com.

Library of Congress Cataloging-in-Publication Data

Names: Roberto, Michael A., author.
Title: Unlocking creativity : how to solve any problem and make the best
  decisions / Michael A. Roberto.
Description: Hoboken, New Jersey : John Wiley & Sons, Inc., [2019] | Includes
  bibliographical references and index. |
Identifiers: LCCN 2018043694 (print) | LCCN 2018045131 (ebook) | ISBN
  9781119545767 (Adobe PDF) | ISBN 9781119545835 (ePub) | ISBN 9781119545798
  (hardcover)
Subjects: LCSH: Creative ability in business. | Creative thinking. | Decision
  making. | Problem solving.
Classification: LCC HD53 (ebook) | LCC HD53 .R596 2019 (print) | DDC
  658.4/03—dc23
LC record available at https://lccn.loc.gov/2018043694

COVER DESIGN: PAUL McCARTHY
COVER IMAGE: © GETTY IMAGES: CLASSEN RAFAEL / EYEEM

Printed in the United States of America

V10006215_111918

To all my teachers from childhood, who stimulated my curiosity and creativity, and most especially to Kristin, the most caring and dedicated elementary school teacher from whom a child could learn.

CONTENTS

PREFACE

Tuesday, September 11, 1979. My family and I sat down in our living room, in front of our Sylvania console television set, at eight o'clock in the evening. We had purchased the *TV Guide* at the supermarket several days earlier to learn about the new fall broadcast network television schedule. Our family had three options that evening. CBS offered a new show, *California Fever,* a rather forgettable teen drama that was canceled after just 10 episodes. NBC televised the debut of *The Misadventures of Sheriff Lobo*, starring Claude Akins.[1] Years later, *TV Guide* ranked that program among the 50 worst television shows of all time.[2] We didn't even give these two programs a second thought. Tuesday evenings served as appointment television in our home. As a nine-year-old boy, I loved watching Arthur Fonzarelli, Richie Cunningham, and the rest of the *Happy Days* crew. Of course, we didn't have much choice. Who in their right mind would watch those other two programs?

Shortly thereafter, my parents signed up for a new service called cable television. I remember the installer bringing a set-top box to our living room. Instead of turning the knob on the front of the television set, we now pushed down one of the box's 12 buttons to change the channel. Imagine that! We now had 12 options instead of just 3! Of course, we still had to get up from the couch each time we wanted to change the channel. The 12th button, the Entertainment and Sports Programming Network (ESPN), proved particularly intriguing to me. No one in my neighborhood had heard of this channel. It promised 24 hours of sports coverage. Mostly, that meant a healthy dose of rodeo, billiards, and Australian-rules football along with college sports (often televised on tape delay!). ESPN did not have the rights to air the major professional sports in those early days. My friends and I mostly loved watching SportsCenter each morning, a show featuring the highlights from the previous day's sporting events, hosted by anchors Bob Ley, George Grande, Tom Mees, and Chris Berman.

Fast-forward 18 years. Cable television had grown considerably, and we had many channels from which to choose. Disney now owned ESPN, and the network aired in over 70 million homes across the country.[3] But in 1997, three important events began to reshape the television landscape. HBO aired its first hour-long original drama (*Oz*), soon to be followed by other groundbreaking and critically acclaimed programs such as *The Sopranos*.[4] Meanwhile, Reed Hastings and Marc Randolph founded Netflix in Scotts Valley, California. The new company offered DVD rentals by mail.[5] Fellow Silicon Valley entrepreneurs Jim Barton and Mike Ramsay founded TiVo in that same year. Their digital video recorder enabled people to record programs, pause live television, and skip commercials easily.[6]

Today the television industry has changed dramatically. Broadcast television viewership has declined substantially over the past two decades. Netflix and Hulu have roughly 75 million subscribers combined in the United States.[7] Cord cutting has become quite prevalent, meaning that more and more consumers choose to go without a cable television subscription. As a result, ESPN has shed 16 million subscribers over the past seven years, which amounts to over $1 billion annually in lost revenue.[8]

Today, my family has an incredible array of high-quality programming options from which to choose. We can select from nearly 500 original scripted programs, up from 182 shows just 15 years ago.[9] On any given day, my children might be binge-watching the new season of a Netflix or Amazon original show in a matter of days, or plowing through every season of old favorites such as *Friends* or *The Office*. My spouse and I could be binge-watching our favorite new program, *The Crown*, while DVRing something that we simply don't have time to view at the moment. Despite the radical change in consumer viewing habits, the broadcast television networks continue to premiere most series in September, air episodes once per week, and televise a season finale in May. Talk about sticking with the status quo.

The past year's Emmy nominations demonstrate how new players dominate the production of high-quality, creative programming. In the best-comedy category, ABC received two nominations, but the rest went to shows airing on HBO, Netflix, and FX. In the best-drama category, only one broadcast network show received a nomination (NBC's *This is Us*). Netflix, Hulu, HBO, and AMC received the other six nods.[10] The last broadcast network program to win this Emmy award was Fox's *24*—and that was 12 years ago.[11]

The transformation of television during my lifetime raises some interesting questions for me. First, how are new companies producing so much highly creative content, and how have they developed new business models? Second, what has prevented traditional players from adapting successfully? Surely, the traditional television players do not lack creative talent. What, then, are the obstacles that prevent them from adopting new business models or generating high-quality, creative content to compete successfully with the likes of Netflix and HBO?

These questions can be generalized and applied across a range of industries and situations. The desire for more creative solutions to pressing problems extends well beyond the television business, of course. When surveyed, CEOs across a variety of industries have identified creativity as one of the most desired leadership qualities for the future.[12] Many companies face a growth crisis, or they find their industries are being disrupted by entrants with different business models featuring original products and services that address unmet consumer needs. These established firms desperately need creative solutions. They must adapt or die.

Seven years ago, with this challenge in mind, my colleagues and I concluded that we needed to enhance the creative capabilities of our students. We had to prepare them better for a changing workplace and turbulent environment. Our team did not believe that creative capacity was a fixed trait. Instead, we embraced the notion that creative capabilities could be nurtured. Our team developed the IDEA program at Bryant University. Every first-year student takes part in this immersive, three-day experience that provides hands-on experience with the design thinking process, a creative problem-solving methodology used by many enterprises. The students' ability to generate breakthrough solutions to perplexing problems in a matter of days always amazes us. We find the experience both exhilarating and inspirational.

As we developed and delivered this unique program over the past seven years, I also spent time researching creativity in organizations around the world. Unfortunately, I have witnessed many impediments to creativity in these enterprises. Senior leaders routinely speak about the need and desire for more creative ideas, but their employees seem frustrated and discouraged when they pose original concepts and solutions. This observation motivated me to write this book. I wanted to understand the barriers to creativity in more depth.

Numerous explanations exist for why organizations fail to generate a sufficient number of creative ideas. One theory focuses on the dearth of talent. In other words, older, established firms simply need to do a better job of attracting and retaining highly creative individuals. Another theory focuses on organizational structure, emphasizing how hierarchy and bureaucracy stifles creativity in many enterprises. Still others attribute the lack of creativity to the pressure to meet Wall Street earnings expectations, or the use of short-term incentive and compensation schemes. These explanations do not lack merit, but they don't tell a complete story.

This book addresses a more fundamental obstacle to creativity in organizations. I examine the organizational mindsets that stifle creativity. By mindsets, I mean the belief systems that shape how people think, decide, and act with regard to the development of original ideas. These mindsets often are quite pervasive, reaching all corners of an organization. They do not reside simply in the heads of a few individuals. The mindsets comprise implicit and explicit beliefs about how the creative process unfolds, what drives creativity, and how creative ideas should be evaluated.

In the chapters that follow, I argue that leaders at all levels need to transform these mindsets to stimulate creativity in their organizations. They should not focus simply on finding "better" people, but instead remove the obstacles that impede the creativity of the talented individuals already in their midst. The best leaders acknowledge that they might not have the creative solutions to their organization's most significant challenges. They seek to marshal the collective intellect of their people and unleash the creative capabilities of those around them. These leaders embrace the responsibility to create a supportive environment and dismantle the barriers to creativity. This book aims to help leaders in this mission to build more creative enterprises. As you read about the six mindsets described in the pages that follow, consider how they shape and influence thought and action in your organization. No matter your position or authority, you can contribute to the successful transformation of these mindsets. One person cannot do it alone. Leaders at all levels, including those without formal authority, will need to partake in this important work.

Michael A. Roberto
June 2018

Endnotes

1. Classic TV Database. "1979–1980 TV Schedule" (www.classic-tv.com/features/schedules/1979-1980-tv-schedule, accessed June 22, 2018).

2. Internet Movie Database. "The 50 Worst TV Shows of All Time According to TV Guide," March 24, 2016 (www.imdb.com/list/ls032245551/, accessed June 22, 2018).

3. Dave Nagle, "ESPN, Inc.: 1997 in Review," ESPN Media Zone, January 2, 1998 (espnmediazone.com/us/press-releases/1998/01/espn-inc-1997-in-review/, accessed June 25, 2018).

4. Ethan Alter, "Return to 'Oz': An Oral History of the Pioneering Prison Drama," Yahoo! TV, July 12, 2017 (www.yahoo.com/entertainment/hbo-oz-20th-anniversary-oral-history-153416770.html, accessed June 25, 2018).

5. Netflix. "About Netflix," Netflix Media Center (media.netflix.com/en/about-netflix, accessed June 25, 2018).

6. TiVo. "History," TiVo.com (www.tivo.com/history, accessed June 25, 2018).

7. Chris Welch, "Hulu Passes 20 Million US Subscribers, Says Offline Downloads Are Coming," The Verge, May 2, 2018 (www.theverge.com/2018/5/2/17309336/hulu-20-million-subscribers-announced-offline-downloads-new-feature, accessed June 26, 2018).

8. Shalini Ramachandran, "How a Weakened ESPN Became Consumed by Politics," *Wall Street Journal*, May 24, 2018 (www.wsj.com/articles/how-a-weakened-espn-became-consumed-by-politics-1527176425, accessed June 26, 2018).

9. John Koblin, "487 Original Programs Aired in 2017. Bet You Didn't Watch Them All," *New York Times*, January 5, 2018 (www.nytimes.com/2018/01/05/business/media/487-original-programs-aired-in-2017.html, accessed June 26, 2018).

10. variety.com/2017/tv/news/2017-emmy-nominees-list-nominations-1202494465/, accessed June 26, 2018.

11. Variety Staff. "Emmys 2017: Full List of Nominations," Variety.com, July 13, 2017 (www.emmys.com/awards/nominees-winners, accessed June 26, 2018).

12. IBM. "IBM 2010 Global CEO Study: Creativity Selected as Most Crucial Factor for Future Success," IBM press release, May 18, 2018 (www-03.ibm.com/press/us/en/pressrelease/31670.wss, accessed June 26, 2018).

CHAPTER 1

The Resistance
to New Ideas

The difficulty lies not so much in developing new ideas as in escaping from old ones.

         —John Maynard Keynes, economist

Many critics rendered harsh judgment when 40-year-old Édouard Manet displayed his rather shocking painting *Le Bain* at an exhibition in Paris on May 15, 1863. Critics responded:

- Its garish colouring pierces the eyes like a steel saw; his figures seem to have been cut out with a punch and have a hardness that is capable of no soothing compromise. It has all the unpalatability of green fruits that will never ripen.[1]
- A young man's practical joke, a shameful open sore not worth exhibiting this way.[2]
- An absurd composition.[3]

Manet's controversial work featured a naked woman seated on the ground alongside two men fully clothed in stylish attire. The woman's blue dress and straw hat lay on the ground beside her, adjacent to a picnic basket and a loaf of bread. In the background, another woman bathes in a stream. Manet's work proved scandalous. He had not depicted a nude goddess in a scene from mythology, as many traditional painters

1

did, but rather an unclothed woman in a modern Parisian scene. Some suggested that the painting depicted prostitutes working in the Bois de Boulogne, a large public park on the western edge of Paris. The painting elicited derision and ridicule from those who attended the exhibition. One person wrote that Manet's work met with a "veritable clamor of condemnation."[4] Another critic observed that, "Never was such insane laughter better deserved."[5]

Le Bain (later retitled *Luncheon on the Grass*) elicited criticism not only due to the scandalous nature of the Parisian scene Manet depicted. It also challenged convention and tradition with its style; many considered Manet's approach quite radical and rather crude. He did not try to capture every detail with precision. Author Ross King wrote that, "[Manet] did not concern himself with realistically transcribing nature or ensuring the flesh tones of his subjects correctly matched their outdoor setting."[6] Instead, *Le Bain* appeared "sketch-like" and "roughly-painted."[7] Manet did not apply his paint in layers over the course of many weeks or even months, and he did not apply a glaze to the finished artwork. Instead, he pioneered the *alla prima* (at once) technique, using broad brushstrokes to paint a scene in one sitting. His work featured sharp contrasts of color rather than subtle transitions. The painting lacked proper perspective, too.[8] Many critics rejected this radical new style. Manet lacked the finesse to which they had become accustomed.

In 1863, many people regarded Jean-Louis-Ernest Meissonier as "the most renowned artist of our time."[9] Unlike Manet, Meissonier worked with great precision to depict scenes of 17th- and 18th-century life, as many other artists did at the time. His work evoked nostalgia for the past, depicting chivalrous gentlemen on horseback or men engaged in noble activities such as chess, music, painting, or reading. Meissonier also loved to depict famous scenes from Napoleon's military campaigns. He strove for historical accuracy and authenticity in every detail. Observers needed a magnifying glass to truly appreciate the minute details captured meticulously in each painting. Critics marveled at his physical dexterity. Meissonier amassed a considerable fortune and received great acclaim for his work. While Meissonier received praise, Manet once noted that, "Insults are pouring down on me as thick as hail."[10]

In that era, French artists aspired to display their work at the Exhibition of Living Artists that took place annually in the Grand Palais des Champs-Élysées. Commonly referred to as the Paris Salon, the exhibition attracted as many as one million citizens over a six-week period. Manet submitted *Le Bain* in 1863, hoping it would be chosen by the members of the jury for inclusion in that year's salon. Count Alfred Émilien O'Hara van Nieuwerkerke oversaw the selection process. He strove to preserve the highest possible standards for the salon. He favored the style of Meissonier, with its focus on history and idealism, and rejected the realism movement, with its embrace of ordinary life and people of all social classes. Commenting on these radical new artists, he said, "This is the painting of democrats, of men who don't change their underwear."[11]

Nieuwerkerke ruled that the jury should consist only of men who were members of the Académie des Beaux-Arts, an elite society of traditionalists intent on preserving the status quo. Approximately, 3,000 artists submitted more than 5,000 paintings for consideration in 1863. In mid-April, the jury announced its decisions. They had accepted only 2,217 paintings by 988 artists. The jury rejected *Le Bain* as well as two other paintings submitted by Manet. Other spurned artists included Gustave Courbet, Pierre-Auguste Renoir, Camille Pissarro, Paul Cézanne, and James Abbott McNeill Whistler. Controversy swirled around the widespread rejections. Emperor Napoleon decided to intervene. Concerned about societal unrest and discontent, the emperor chose to embrace the idea of a separate exhibition consisting of the artwork rejected by the establishment. Soon this exhibition came to be known as the Salon des Refuses (exhibition of the rejects). More than 1,000 people per day attended, though many laughed at the rejected works of art. Manet submitted *Le Bain* for display, and mockery and ridicule ensued for him as well.

Amidst the deluge of criticism, a few astute observers noted the stark contrast between those accepted and rejected by the Paris Salon. They sensed that the ground had begun to shift. The famous journalist and art critic Théophile Thoré described it as a contrast between "conservatives and innovators, tradition and originality."[12] Amidst widespread criticism, younger artists took comfort that others shared their willingness to experiment and break new ground. Manet became a leader among this new generation of painters. He met regularly with other innovators such as Edgar Degas, Claude Monet, Renoir, and Pissarro at Café Guerbois in Paris. They argued and debated, and they shared ideas on Sundays and Thursdays, becoming known as the Batignolles Group.

Ten years after the original salon controversy, Monet, Renoir, Pissarro, Degas, and others created the Société Anonyme Coopérative des Artistes Peintres, Sculpteurs, Graveurs (Cooperative and Anonymous Association of Painters, Sculptors, and Engravers). They chose not to submit their work to the Paris Salon. Instead, they formed an independent exhibition, which opened to mixed reviews. Monet submitted a painting titled, *Impression, Sunrise.* Critic Louis Leroy mocked the painting in an article titled, *The Exhibition of the Impressionists.* He wrote, "Impression—I was certain of it. I was just telling myself that, since I was impressed, there had to be some impression in it . . . and what freedom, what ease of workmanship! Wallpaper in its embryonic state is more finished than that seascape."[13] Others started referring to this group of renegade artists as the *impressionists,* and even the painters themselves adopted the name despite the fact that it had emerged from a scathing criticism of their work. We know how this story ends. Ultimately, Manet became known as the father of modernism, and the impressionist movement stands as one of the most consequential eras in art history.

The story of Manet and the impressionists should not surprise us. We have heard this type of story on many occasions. Today's experts reject tomorrow's creative geniuses. Conventional wisdom, preconceived notions, and cognitive biases blind the experts from recognizing the merits of bold new ideas. We trust experts and look to them for wise judgment, prescient forecasts, and sound leadership. Turn on the television, and you see a steady stream of pundits being called upon to weigh in on a variety of economic, political, and social issues. However, expertise may not translate into an ability to see the future, or to evaluate original, out-of-the-box ideas more effectively than you and I can. Experts should be flying aircraft, performing heart surgeries, and designing bridges. We don't want a novice fixing our car or our broken hip. However, when it comes to creativity and innovation, expertise may be a liability at times. As Zen teacher Shunryu Suzuki once said, "In the beginner's mind there are many possibilities, in the expert's mind there are few."[14]

Closed-Minded Experts

Alfred Wegener brought a beginner's mindset to the field of geology over a century ago. Like Manet, his fresh ideas did not earn acceptance readily. Wegener earned a doctorate in astronomy in 1904 and later became immersed in meteorological research. He became fascinated by the discovery of similar animal and plant organisms on different continents, as well as complementary geological features on landmasses separated by oceans. He proposed his theory of continental drift in the early 1900s. Geologists forcefully rejected his ideas. Rollin T. Chamberlin of the University of Chicago commented, "Wegener's hypothesis in general is of the footloose type, in that it takes considerable liberty with our globe, and is less bound by restrictions or tied down by awkward, ugly facts than most of its rival theories."[15] Wegener's concept only became widely accepted by scientists decades after his death.

Chester Carlson invented the process of electrophotography in the 1930s, but many companies rejected his requests for funding. Writing years later, Harold Clark noted that:

> Xerography had practically no foundation in previous scientific work. Chet put together a rather odd lot of phenomena, each of which was obscure in itself and none of which had previously been related in anyone's thinking. The result was the biggest thing in imaging since the coming of photography itself.[16]

Finally, in the mid-1940s, the company later known as Xerox decided to support Carlson. By 1965, the Xerox 914 copier accounted for over $240 million in revenue, over 60 percent of the company's total revenue. The word *Xerox* became a verb, much like *Google* is today.

In the 1980s, Barry Marshall and Robin Warren argued that bacterial infections, rather than stress, caused ulcers. Marshall explained the initial reception when he began presenting his work at medical conferences:

> To gastroenterologists, the concept of a germ causing ulcers was like saying that the Earth is flat. After that I realized my paper was going to have difficulty being accepted. You think, "It's science; it's got to be accepted." But it's not an absolute given. The idea was too weird.[17]

Frustrated by the mainstream medical community's reaction to his work, Marshall took some *Helicobacter pylori* bacteria from the stomach of an ailing patient, ingested it himself, and became quite ill. Within days, Marshall experienced vomiting, halitosis, and gastritis (an inflammation of the stomach lining). He treated himself with antibiotics and he recovered fully. Still, experts did not accept Marshall and Warren's theory for years. Finally, in 2005, they received the Nobel Prize in Medicine for their groundbreaking work.

We always hear the stories of venture capitalists striking it rich by investing at the ground level in startups that go on to achieve remarkable success. For instance, Peter Thiel invested $500,000 in Facebook in 2004. Eight years later, he sold his stake in the social media giant for more than $1 billion. However, many entrepreneurs face multiple rounds of rejection by industry experts. For example, Joe Gebbia, Brian Chesky, and Nathan Blecharczyk sought funding for their startup in 2008. They wanted to raise $150,000 in return for a 10 percent stake in their company. The co-founders approached seven accomplished and well-known investors in Silicon Valley. Five investors sent them rejection letters, while two never even replied.[18] Nine years later, their company, Airbnb, had achieved a $31 billion valuation. If one of these investors had invested back in 2008, their $150,000 investment would have been worth $3.1 billion nine years later. The Airbnb story does not prove to be unique. Even the most accomplished venture capitalists invest in many startups that do not succeed and pass on a number of deals that could have been highly lucrative. Every investor has at least one great regret.

Erin Scott, Pian Shu, and Roman Lubynsky examined data on 652 startups from MIT's Venture Mentoring Service. The service attempts to match startups with mentors. The mentors receive data about a variety of startup ideas. They must decide what they think about the ideas without having an opportunity to review information about the founders or to meet the team in person. Scott and her colleagues examined how many of these startups went on to have their products commercialized successfully. For startups involving high research and development expenditures, the more highly rated ideas did have a better chance of being commercialized. However, the researchers checked to see if expert mentors were better at predicting a startup's success than the mentor group overall. They defined experts as people with industry-specific experience or doctoral degrees in that particular technical field. The study's results suggest that expert mentors with extensive industry experience and academic training did not forecast new venture success in R&D-intensive sectors more accurately than the mentor group overall.[19]

Why do experts fail to recognize creative genius? Victor Ottati and his colleagues have documented evidence of what they call the *earned dogmatism effect*. The scholars argue that social norms about novices versus experts play a key role in how people perceive new ideas. They explain as follows:

> Consider, for example, a seminar pertaining to cancer. Within this situation, some individuals may occupy the role of "novice" (e.g., a layperson) whereas others may occupy the role of "expert" (e.g., a cancer researcher). Because novices possess limited knowledge, social norms dictate that they should listen and learn in an open-minded fashion. The expert possesses extensive knowledge, and therefore is entitled to adopt a more dogmatic or forceful orientation. Dogmatic statements are more likely to be tolerated when the "expert" speaks than when a "novice" speaks. Novices possess limited knowledge, and as such, are expected to adopt a more humble and open-minded orientation.[20]

Ottati and his co-authors conducted a series of six experiments to study the earned dogmatism effect. In particular, they wanted to know if self-perceptions of expertise mattered. In other words, does close-minded behavior occur simply because people *perceive themselves* to be experts, even if that might not actually be the case? In the studies, individuals were made to feel as though they were either experts or novices in a particular knowledge domain. The scholars discovered that those who felt as though they were experts tended to act in a more close-minded fashion in subsequent parts of the study. For instance, they gave a political history test to research subjects in one experiment (15 multiple-choice questions such as "Who was Richard Nixon's initial vice president?"). One-half of the subjects received easy questions, while the others tackled challenging questions. After the participants responded to all the questions, the researchers provided them false feedback. They told participants who had answered the easy questions that they had performed better than 86 percent of the test-takers. They informed the subjects responding to difficult questions that they had performed very poorly, worse than 86 percent of their fellow test-takers. The scholars then administered a cognitive test of open-mindedness. The participants who had received the positive feedback (made to feel as though they were experts) tended to exhibit more closed-mindedness, even though the feedback was completely made up! Ottati and his co-authors concluded that people become more dogmatic when they perceive themselves as experts.

Dogmatic thinking and closed-mindedness may be most prevalent when outsiders or newcomers propose theories that mark a radical break from past convention. In 1962 Thomas Kuhn wrote a groundbreaking book titled *The Structure of Scientific Revolutions*. A physicist by training, Kuhn became one of the most influential philosophers and historians of science. He argued that science does not progress solely in a linear, incremental, and evolutionary fashion. Instead, major leaps forward occur from time to time in a revolutionary fashion. Kuhn describes these discontinuities as paradigm shifts. Controversial new models shake the foundation of a field during these revolutions. Kuhn argued that newcomers often drive the paradigm shifts:

> Almost always the men who achieve these fundamental inventions of a new paradigm have been either very young or very new to the field whose paradigm they change. And perhaps that point need not have been made explicit, for obviously these are the men who,

being little committed by prior practice to the traditional rules of normal science, are particularly likely to see that those rules no longer define a playable game and to conceive another set that can replace them.[21]

The story of Wegener represents one of those groundbreaking paradigm shifts triggered by a newcomer. Experts did not simply reject his ideas about continental drift because they challenged the prevailing paradigm. Undoubtedly, the resistance to his ideas existed, in part, because he had not been trained as a geologist. Outsiders often drive paradigm shifts because they do not exhibit a bias toward the status quo. However, their outsider status and lack of specialized training makes it difficult for them to gain acceptance for their theories. How can an astronomer and meteorologist overturn centuries of thought in the field of geology? It's simply not possible! Max Planck, winner of the Nobel Prize in Physics in 1918, once remarked on the challenge of overturning a scientific paradigm. He commented, "A new scientific truth does not triumph by convincing its opponents and making them see the light, but rather because its opponents eventually die, and a new generation grows up that is familiar with it."[22]

Double Talk on Creativity

Many creative individuals working in corporations today encounter the same type of resistance that trailblazing artists, scientists, and inventors have experienced throughout history. Experts reject their ideas and prefer to defend the status quo. Technical specialists exhibit closed-minded behavior when newcomers challenge the conventional wisdom or question established practices. Newcomers experience pressures for conformity. Leaders create an environment where people with new ideas fear speaking up. The organizational culture does not promote experimentation and risk-taking behavior. Rewards and incentive systems focus on efficiency and productivity, and they discourage learning and exploration.

At the same time, corporate leaders speak often about the need for creativity and innovation. They claim that creativity has become their

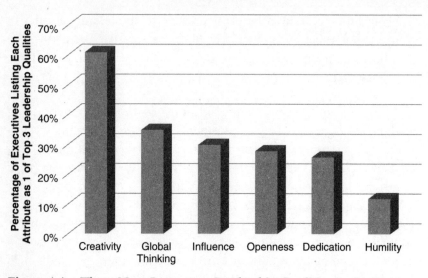

Figure 1.1   Three Most Important Leadership Qualities in the Next Five Years

Source: Data extracted from 2010 IBM CEO Study.[24]

highest priority. Several years ago, IBM conducted a Global CEO Study. The technology giant surveyed 1,541 chief executives, general managers, and public-sector leaders across 33 industries and 60 countries around the globe. Approximately 60 percent of these executives cited creativity as the most important leadership attribute needed for future success (see Figure 1.1).[23]

Chief executives say the right thing when it comes to creativity, but do they walk the walk? In 2016 the O.C. Tanner Institute surveyed approximately 3,500 employees from firms in five countries around the world. They found that most employees did not feel supported and inspired by their leaders. While executives named innovation as a top priority, most employees felt that they were not encouraged to develop new ideas, and they did not have the time and resources required to do so. The O.C. Tanner Institute concluded that many employees become disheartened and cynical when they perceive leadership calls for more creativity and innovation as "largely empty talk."[25]

Creativity Bias

Unfortunately, a bias against creativity may be quite prevalent in organizations. Consider the work of scholar Jennifer Mueller at the University of San Diego. Her research shows that people have decidedly mixed feelings about creativity. She argues that social norms may cause people to express positive attitudes about creativity. However, people's actual behaviors may not be consistent with their espoused beliefs. She and her colleagues conducted experiments demonstrating that people value practicality over creativity when faced with conditions of uncertainty and ambiguity. Moreover, individuals may be less able to recognize creative ideas when motivated to reduce uncertainty.[26]

Mueller finds that creative individuals also may face a "penalty" when it comes to others assessing their leadership potential. She and her colleagues conducted three studies that showed that people have a tendency to view creative individuals as having less leadership potential (unless they are told to focus on charismatic individuals). Why might that be? The authors argue that people tend to have mixed feelings about creative colleagues.[27] Mueller explains, "In addition to 'visionary' and 'charismatic,' people also use words like 'quirky,' 'unfocused' [and] 'nonconformist.' The fact is people don't feel just positively about creative individuals—they feel ambivalent about them."[28] The research demonstrates a powerful dilemma. On the one hand, people express a strong desire to have creativity as a characteristic of their leaders. On the other hand, when people offer out-of-the-box ideas, they sometimes are viewed in a negative light. People think that these inventive colleagues are strange or perplexing.

The bias against creativity does not reside only in corporations. Our schools may be discouraging creative students in a variety of ways. A stream of research has shown that teachers claim to value qualities such as independent thinking and curiosity, yet they reward behaviors such as obedience and conformity.[29] Ken Robinson has been one of the most outspoken critics of how we educate our children. He quotes Picasso, who once said, "Every child is an artist. The problem is how to remain an artist once he grows up." Robinson argues that our schools place a premium on certain forms of intelligence and ability, namely those associated with traditional academic ability. Robinson jokes, "The whole purpose of public education throughout the world is to produce university professors."[30] Moreover, teachers and parents often steer students toward fields of study that they believe will lead most readily to steady employment. Look at all the attention placed on STEM disciplines these days. As a society, we seem to be saying to our children that engineering is valuable and worthwhile to study, while music or art is not.

As we grow up, we adopt attitudes and mindsets that inhibit our creativity in many ways. Consider an exercise developed years ago by Stanford creativity researcher Bob McKim. He asked adults to spend 30 seconds drawing a sketch of their neighbor. Imagine the murmurs in the room. People often squirm in their seats when presented with this task. The discomfort appears palpable. After 30 seconds, McKim asks people to show the sketch to their neighbor. What typically happens? People display their sketches to others with great trepidation. They apologize profusely for their terrible drawing. They express embarrassment at their lack of artistic ability. Interestingly, McKim found that children react quite differently to this same task. They embrace it with enthusiasm. They demonstrate pride of authorship when showing their neighbors the sketches.[31] My experience confirms McKim's finding. My 10-year old son beams with pride when he brings home an art project from school. When I'm asked to sketch an idea during a brainstorming session, I often cringe. What happens as we grow up? For a variety of reasons, we fear how others will perceive or judge us. That mindset of fear impedes our ability to be creative. Could our schools be contributing to this mindset of fear and this attitude of risk aversion? Robinson certainly believes so. Perhaps renowned science fiction author and biochemist Isaac Asimov was right when he said, "The world in general disapproves of creativity, and to be creative in public is particularly bad. Even to speculate in public is rather worrisome."[32]

The Dire Need for Creativity

We need creativity more than ever though. We have perplexing problems to solve in education, healthcare, and poverty. In business, established companies in a variety of industries face serious threats to their survival. Traditional brick-and-mortar retailers, for instance, struggle to cope with the e-commerce revolution. Firms such as Toys R Us, Sports Authority, and Radio Shack have gone bankrupt in recent years. One of the most storied retailers of the twentieth century, Sears, entered bankruptcy as well in October 2018 after years of decline.

Chief executives claim that creativity is a top priority because they desperately seek growth and renewal. Many large companies face a growth crisis today. A recent analysis (see Figure 1.2) demonstrates that more than one-third of the firms on the Fortune 500 list experienced a decline in revenue from 2014 to 2016. Sales declined by more than 10 percent for 59 of these companies. Meanwhile, many firms achieved slow single-digit revenue growth during this period.[33]

Investors, of course, value growth a great deal. Take the case of Amazon. While the firm has generated very little profit during its 23-year history, shareholder returns have been tremendous. Imagine that you had invested $5,000 in Amazon at its initial public offering. Precisely 20 years later, your investment would have been

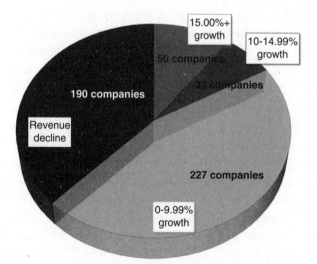

Figure 1.2   Fortune 500 Compound Annual Growth Rate Analysis: 2014–2016

Source: Analysis based on data compiled by Craft[34]

worth $2.4 million.[35] The Amazon story is not unique. McKinsey and Company conducted a study of approximately 3,000 software and online-services companies over a 22-year period. They discovered that the fastest-growing companies generated shareholder returns five times greater than the average firm. What happens when growth stalls? Few companies recover from a lengthy period of stagnation or decline, and not many chief executives keep their jobs in these situations. Employees at all levels suffer, as layoffs and facility closures often take place and morale plummets. Massive cost cutting might prop up earnings for a short time, but ultimately, firms cannot survive if the top line keeps shrinking.[36]

Companies desperately desire new growth engines, but unfortunately, it appears that great new ideas are becoming harder to find. Nicholas Bloom and his colleagues examined data on research productivity for the economy as a whole, as well as in specific industries such as semiconductors, agriculture, and medicine. They find that amazing discoveries and inventions continue to occur, but that's because we have devoted more people and resources to our innovation efforts. Meanwhile, research productivity has fallen significantly in recent decades.

In 1965 Intel co-founder Gordon Moore observed that the number of transistors per square inch of integrated circuit had doubled approximately every two years. He predicted that this trend would continue, and he proved correct in the decades that followed. However, the continued advances in semiconductor technology have required increasing amounts of investment. That research effort has been much less productive over time. Bloom and his colleagues find that, "Because of declining research productivity, it is *around 18 times* harder today to generate the exponential growth behind Moore's Law than it was in 1971."[37]

In industry after industry, companies have poured more people and more dollars into research efforts, in hopes of generating new discoveries and new revenue streams. Examining the economy as a whole, Bloom and his colleagues find that, "Since the 1930s, research effort has risen by a factor of 23—an average growth rate of 4.3 percent per year. Research productivity has fallen by an even larger amount—by a factor of 41 (or at an average growth rate of −5.1 percent per year)."[38] When it comes to innovation, we simply are not getting the bang for our buck that we used to achieve.

The Person versus the Situation

How can we enhance the pace of invention and discovery? How do we find more creative ideas to solve our most perplexing problems? We can throw more resources at this challenge, as we have been doing for decades. Or, perhaps large companies haven't been hiring the right people. If we just find the right people, the truly breakthrough thinkers, perhaps new ideas will flourish. *Inc.* magazine ran a special feature several years ago about hiring for creativity. The writers argued that companies have to do a better job of recruiting individuals who think outside the box.[39] Indeed, some firms have taken to using a battery of tests to evaluate the creativity of job candidates. According to this line of thinking, innovative companies such as Netflix, Amazon, and Google simply have done a better job than most firms at finding the most creative minds.

Step back for a moment though. Consider the story of forty-seven students at Princeton Theological Seminary in December 1970. Two professors asked them to prepare and record a 3-to-5-minute talk as part of a study about the careers of seminary students. One-half of the students prepared talks regarding the range of jobs or professions in which seminary students could be effective and content. The others prepared talks about the Parable of the Good Samaritan (a story from the Bible about the choice to neglect or help an injured stranger along the road). After several minutes, the researchers informed the students that they would have walk to a different campus building to record their talks. They told some students to hurry over to the other building, while informing others that they had ample time and might even have to wait before recording their speech.

When the theology students walked across campus, they encountered someone slumped in a doorway with their eyes closed and their head down. The "victim" coughed and groaned when a student approached. The researchers recorded the extent to which each student offered assistance to the victim. What did they find? Not surprisingly, the students in a hurry tended to offer much less assistance to the person slumped in the doorway. Perhaps more shockingly, the students who had prepared the talk about the Good Samaritan were no more likely to lend a hand than those who had prepared the generic speech about future jobs. Several students actually stepped over the victim as they rushed to the next building to present their lecture! Differences in personality type did not explain why some helped the victim, while others did not.[40]

Our initial inclination might be that "good" people help others, while "bad" people rush past. We could not imagine ourselves ignoring a fellow human being lying injured in a doorway. However, this study suggests that we underestimate the power of circumstances. The situation and our environment shape our behavior more than we would like to think. Time and again, our actions reinforce this fundamental point about human behavior.

Ask yourself: Did Paris in the 1850s lack creative new artists or did the salon system squelch the emergence of new techniques and perspectives? All leaders in search of fresh ideas must ask themselves whether they have a people problem or a situation problem. Do firms have a dearth of creative people, or are they blocking and resisting the Manets, Wegeners, and Marshalls already in their midst? If leaders conclude that they have a people problem, they will spend inordinate amounts of time and energy trying to find more creative job candidates. They may not spend much time looking in the mirror, examining their own closed-minded or dogmatic behavior, or evaluating the culture they have created. They might not evaluate how their organization treats newcomers with fresh perspectives.

Alternatively, leaders might come to recognize that all people have creative potential. We just need the right conditions in which to flourish. For many growth-starved firms, the problem resides in the environment, not the workforce. The best leaders identify how and why their organizations may resist new ideas or exhibit a creativity bias. They build environments that neither marginalize those who challenge the conventional wisdom nor punish those who often refuse to conform. Once leaders recognize that they have a situation problem, not a people problem, they can begin identifying and removing the true barriers to creativity in their organizations.

The Six Mindsets

In the pages that follow, we will explore how leaders can reshape the organizational environment so as to enable creativity to blossom.

Specifically, the book describes the six organizational mindsets that must be transformed for people to fulfill their creative potential. These mindsets encompass a collection of explicit and implicit beliefs that shape how people analyze and evaluate, make decisions, and take action with regard to imaginative, original ideas. These belief systems may permeate all facets of an enterprise. The six creativity-inhibiting mindsets are:

- **The Linear Mindset:** Many organizations fail to understand and embrace the iterative and discontinuous nature of the creative process. They mistakenly try to move from analysis to idea formulation to execution in a step-by-step manner.
- **The Benchmarking Mindset:** Firms study their competitors closely, but in so doing, they experience fixation. Consequently, they adopt copycat approaches rather than creating distinctive strategies.
- **The Prediction Mindset:** Managers have a desperate desire to see what's next and they exhibit overconfidence in the ability of experts to forecast the future. The insatiable need to predict just how big ideas will become actually impedes creativity.
- **The Structural Mindset:** Managers often resort to changes in organizational structure as a means of stimulating creativity and improving performance. They fail to recognize the limited efficacy of redrawing the lines and arrows on the organization chart time and again.
- **The Focus Mindset:** Organizations believe that teams will excel at creative work if they focus intensively, perhaps even secluded from their colleagues. They fail to recognize that the best creative thinkers oscillate between states of focus and unfocus.
- **The Naysayer Mindset:** Managers encourage people to critique each other's ideas early and often. Unfortunately, the failure to manage dissent and contrarian perspectives constructively causes many good ideas to wither on the vine.

These mindsets represent powerful obstacles that must be dismantled for the creative process to thrive. Leaders do not need to generate more great ideas. They must clear the path so that curious thinkers throughout their teams and organizations can experiment, learn, and discover. Trust your people, remove the hurdles, and bold and original ideas will come forth.

Endnotes

1. Ernest-Alfred Chesneau qtd. in "Dejuener sur l'herbe (1863) by Edouard Manet," Encyclopedia of Art Education (www.visual-arts-cork.com/ paintings-analysis/luncheon-on-the-grass.htm, accessed October 16, 2017).

2. Louis Étienne qtd. in Ross King, *The Judgment of Paris: The Revolutionary Decade that Gave the World Impressionism* (New York: Bloomsbury, 2007), 88.

3. Théophile Thoré qtd. in King, *The Judgment of Paris*, 89.

4. King, *The Judgment of Paris*, 87.

5. King, *The Judgment of Paris*, 88.

6. King, *The Judgment of Paris*, 39.

7. Anahita Shafa, "Manet," Impressionist Impressions website, 2007 (www.anahita design.com/impressionist/manet.html, accessed October 16, 2017).

8. Musée d'Orsay, "Edouard Manet: *Luncheon on the Grass*," 2018 (www.musee-orsay.fr/index.php?id=851&L=1&tx_commentaire_pi1%5BshowUid%5D= 7123, accessed October 17, 2017).

9. King, *The Judgment of Paris*, 2.

10. King, *The Judgment of Paris*, 151.

11. King, *The Judgment of Paris*, 32.

12. King, *The Judgment of Paris*, 91.

13. C. Monet Gallery, "Monet, Father of Impressionism," (www.cmonetgallery .com/father-of-impressionism.aspx, accessed October 17, 2017).

14. Shunryu Suzuki, *Zen Mind, Beginner's Mind: Informal Talks on Zen Meditation and Practice* (Boston: Shambhala, 2010), 1.

15. James Lawrence Powell, *Mysteries of Terra Firma: The Age and Evolution of the World* (New York: Free Press, 2001), 92.

16. David Owen, *Copies in Seconds: Chester Carlson and the Birth of the Xerox Machine* (New York: Simon and Schuster, 2004), 92.

17. Pamela Weintraub, "The Dr. Who Drank Infectious Broth, Gave Himself an Ulcer, and Solved a Medical Mystery," *Discover*, March 2010 (discovermagazine .com/2010/mar/07-dr-drank-broth-gave-ulcer-solved-medical-mystery, accessed October 18, 2017).

18. Alice Truong, "The rejection letters of early-round investors who passed on Airbnb," Quartz.com, July 13, 2015 (qz.com/452185/the-rejection-letters-of-early-round-investors-who-passed-on-airbnb/, accessed October 19, 2017).

19. Erin Scott, Pian Shu, and Roman M. Lubynsky, "Are 'Better' Ideas More Likely to Succeed? An Empirical Analysis of Startup Evaluation," *Harvard Business School Working Paper,* 16-013, 2016.

20. Victor Ottati, Erika D. Price, Chase Wilson, and Nathanael Sumaktoyo, "When self-perceptions of expertise increase closed-minded cognition: The earned dogmatism effect," *Journal of Experimental Social Psychology*, 61 (2015): 132.

21. Thomas Kuhn, *The Structure of Scientific Revolutions*, 4th edition (Chicago: University of Chicago Press, 2012), 90.

22. Brandon Brown, "Genius Move," *Slate*, June 2015 (www.slate.com/articles/ health_and_science/science/ 2015/06/max_planck_s_principle_physics_and_ constant_he_knew_how_to_change_his_mind.html, accessed October 20, 2017).

23. IBM. "IBM 2010 Global CEO Study: Creativity Selected as Most Crucial Factor for Future Success," IBM press release, May 18, 2018 (www-03.ibm.com/press/ us/en/pressrelease/31670.wss, accessed June 26, 2018).

24. IBM, *Capitalizing on Complexity: Insights from the Global Chief Executive Officer Study*, 2010 (public.dhe.ibm.com/common/ssi/ecm/gb/en/gbe03297usen/ GBE03297USEN.PDF, accessed October 16, 2017).

25. David Sturt and Jordan Rogers, "A Global Survey Explains Why Your Employees Don't Innovate," *Harvard Business Review* (digital article), February 24, 2016 (hbr.org/2016/02/why-your-employees-dont-innovate, accessed October 16, 2017).

26. Jennifer S. Mueller, Shimul Melwani, and Jack Goncalo, "The Bias Against Creativity: Why People Desire but Reject Creative Ideas," *Psychological Science*, 23(1), 2011, 13–17.

27. Jennifer Mueller, Jack Goncalo, and Dishan Kamdar, "Recognizing creative leadership: Can creative idea expression negatively relate to perceptions of leadership potential?" *Journal of Experimental Social Psychology*, 47(2), 2011, 494–498.

28. Knowledge@Wharton, "A Bias against 'Quirky'? Why Creative People Can Lose Out on Leadership Positions," Knowledge@Wharton, February 16, 2011 (knowledge.wharton.upenn.edu/article/a-bias-against-quirky-why-creative-people-can-lose-out-on-leadership-positions/, accessed October 21, 2017).

29. Erik Westby and V.L. Dawson, "Creativity: Asset or Burden in the Classroom," *Creativity Research Journal*, 8(1), 2010, 1–10.

30. Ken Robinson, "Do Schools Kill Creativity?," TED2006 Conference, February 2006 (www.ted.com/talks/ken_robinson_says_schools_kill_ creativity, accessed October 5, 2017).

31. Tim Brown, "Tales of Creativity and Play," 2008 Serious Play Conference, May 2008 (www.ted.com/talks/tim_brown_on_creativity_and_play, accessed October 5, 2017).

32. Isaac Asimov and Arthur Obermayer, "Isaac Asimov Asks, 'How Do People Get New Ideas?': A 1959 Essay by Isaac Asimov on Creativity" *MIT Technology*

Review, October 20, 2014 (www.technologyreview.com/s/531911/isaac-asimov-asks-how-do-people-get-new-ideas/, accessed October 21, 2017).

33. Tyler Durden. "The Fortune 500's Fastest Growing (and Shrinking) Companies, ZeroHedge, April 5, 2017 (www.zerohedge.com/news/2017-04-05/fortune-500s-fastest-growing-and-shrinking-companies, accessed March 19, 2018).

34. Ibid.

35. Lucinda Stein, "If You Invested in Amazon at Its IPO, You Would Be a Millionaire Today," *Fortune*, May 15, 2017 (fortune.com/2017/05/15/amazon-stock-20-years-ipo/, Accessed November 3, 2017).

36. Eric Kutcher, Olivia Nottebohm, and Kara Sprague, "Grow Fast or Die Slow," April 2014 (www.mckinsey.com/industries/high-tech/our-insights/grow-fast-or-die-slow, accessed October 24, 2017).

37. Nicholas Bloom, Charles Jones, John Van Reenen, and Michael Webb, "Are Ideas Getting Harder to Find?," *National Bureau of Economic Research Working Paper Series*, September 2017, 18.

38. Ibid, 8.

39. Inc. Staff, "How to Hire for Creativity," *Inc.*, October 1, 2010 (www.inc.com/magazine/20101001/guidebook-how-to-hire-for-creativity.html, accessed October 18, 2017).

40. John Darley and C. Daniel Batson, "'From Jerusalem to Jericho': A Study of Situational and Dispositional Variables in Helping Behavior," *Journal of Personality and Social Psychology,* 27(1), 1973, 100–108.

CHAPTER 2

The Linear Mindset

My process is messy and non-linear, full of false starts, fidgets, and errands that I suddenly need to run now; it is a battle to get something—anything—down on paper.

—Ellen Klages, science fiction author

The painter became distracted quite easily, procrastinated frequently, and abandoned many projects before completion. His tendency for perfectionism caused him always to see faults in his work and to iterate unceasingly. He often stepped away from painting to pursue other interests in a wide range of disciplines. These diversions became all-consuming inquiries on which he could fixate on minute details for lengthy periods of time. One day, you might find him studying how to build a weapon of warfare. Soon thereafter, he might be dissecting a cadaver, trying to understand how the aortic valve functions. All the while, patrons awaited the paintings that they had commissioned. He seemed to enjoy idea generation and the exploration of a new field of study much more than the process of execution. Patrons naturally became frustrated at times with his lack of discipline, and they stopped payments when it became clear that commissioned works would remain unfinished.

That painter, Leonardo da Vinci, always fascinated me. My interest was sparked after several trips to his birthplace in Tuscany. When I was young, my uncle would take me to the small museum there, which was near his home in Italy. There, I first learned that da Vinci was much more than a masterful painter. Who could forget his fanciful drawings of flying machines, which the museum displayed?[1] Leonardo exhibited wide-ranging knowledge and expertise across many disciplines. He embodied the notion of a polymath, or what people commonly refer to today as a Renaissance man. Leonardo did much more than paint masterpieces such as *Mona Lisa*, *Last Supper*, and *Virgin of the Rocks*. He also studied anatomy, architecture, engineering, botany, geology, mathematics, and other topics. Leonardo often studied science to inform his art, as evidenced by how his optics research helped him master the use of perspective and shadows in his paintings. However, he pursued scientific research for broader reasons as well. Leonardo exhibited an insatiable intellectual curiosity, eager to learn a great deal about the world around him. That curiosity would become both a blessing and a curse throughout his lifetime.[2]

At age 14, Leonardo obtained an apprenticeship with Andrea del Verocchio in the city of Florence. The young boy studied and worked in Verocchio's workshop for approximately a decade. He finally set out to work independently in 1477, opening his own workshop. Leonardo spent five years there before moving north to Milan. In 1478, he received his first commission from Lorenzo de' Medici for a painting in the Chapel of St. Bernardo of Florence's Palazzo della Signoria.[3] Though he did some planning for this work, he apparently never began the actual painting. During his five years in Florence, Leonardo obtained two additional commissions for two paintings later considered masterpieces: *The Adoration of the Magi* and *Saint Jerome in the Wilderness*. However, he did not actually complete either one. His perfectionism became a roadblock. His notebooks during this time show evidence of personal anguish, turmoil, and self-doubt, perhaps contributing to his inability to complete these paintings. Biographer Walter Isaacson describes another reason why Leonardo may have abandoned these projects:

> He preferred the conception to the execution. As his father and others knew when they drew up the strict contract for his commission, Leonardo at twenty-nine was more easily distracted by the future than he was focused on the present. He was a genius undisciplined by diligence.[4]

The pattern repeated itself many times during his lifetime. Leonardo received commissions, procrastinated, became engrossed in intellectual diversions, and failed to complete his work. The *Mona Lisa* itself emerged from a fascinating process stretching from 1503 until da Vinci's death in 1517. He began the painting in Florence and he continued to make revisions for years, taking it with him as he moved back to Milan, then Rome, and ultimately to France. When Leonardo died, he left the *Mona Lisa* to his patron and friend, King Francois I, who displayed the painting in his palace at Fontainebleau.

In today's world, an employee such as Leonardo would frustrate many leaders. His inability to fulfill commitments would be maddening, as would the unwillingness to focus exclusively on a designated task. Many organizations might not appreciate his genius because they expect disciplined execution—on time, under budget. As the saying goes, you should underpromise and overdeliver. Steve Jobs once remarked, "Real artists ship." In other words, creative individuals ultimately must stop tinkering and deliver a finished product. Otherwise, they cannot achieve commercial success or have an impact on customers' lives. Da Vinci did not live by this mantra. Though he signed contracts and accepted upfront payments at times, these legal and financial commitments did not compel him to complete projects on time, if at all.

Da Vinci loved to experiment. He did not deal strictly in hypotheticals or abstract theories. Da Vinci recognized that one could resolve a point of uncertainty or examine the validity of an idea by devising a good test. As an example, he posed a hypothesis as to how the aortic valve shut in the human body. To test his idea, he devised an experiment by building a glass model from a wax mold of a bull's heart. He placed grass seeds in water and circulated it through his model, with the seeds enabling him to see precisely how the fluid moved and the valve shut. In so doing, he found support for his hypothesis. It took more than four centuries for doctors to recognize that da Vinci's hypothesis was correct, toppling entrenched views in the medical profession.

Da Vinci used prototyping to hone his ideas and creations. Prototyping enabled him to explore new ideas quickly and test novel techniques. By sketching and building, he could think more clearly. Da Vinci regularly created a series of compositional sketches and preparatory drawings before beginning to paint. At times, he even drew cartoons as part of this prototyping process. When preparing to paint *The Virgin and Child with Saint Anne,* one of his cartoons became a sensation when it was displayed publicly in Florence. Da Vinci employed prototyping in other types of projects as well. Upon receiving a commission for a large equestrian monument in 1489, Leonardo conducted anatomical research on horses. He drew various sketches, and ultimately, he built a clay model of the equestrian statue. Similarly, when working on a design for a cathedral's tower in Milan, he constructed a wooden model to illustrate his most ingenious concepts.

Leonardo embraced a learning-by-doing approach in all creative endeavors. He did not plan and then execute in a linear fashion. Instead, he engaged in an iterative process of preparation, reflection, and adaptation. Leonardo viewed his paintings as journeys, not destinations. He planned and sketched before he began painting, but he adapted his approach and techniques by reflecting upon the work he had done. Leonardo never quite considered a painting complete, because his learning-by-doing approach always led to important refinements. He might put a painting aside for some time, but then a burst of new ideas and discoveries would draw him back to that particular work of art. His scientific research, including many human and animal dissections, informed his art. As Leonardo learned more about anatomy, he would alter the way he painted body parts and depicted movements. As Isaacson wrote, "He knew that there was always more he might learn, new techniques he might master, and further inspirations that might strike him."[5] In today's world, we need to ship eventually, but to get to that point, we would be well served to pursue an iterative process of action and adaptation. Experimentation and prototyping enable us to learn and improve quickly. Even after we ship, we should continue learning so that we might refine our products and services in the future. We are never truly finished.

Linear Thinking

In contrast to Da Vinci's creative process, today's managers often think about problem solving in a linear fashion (see Figure 2.1). First plan

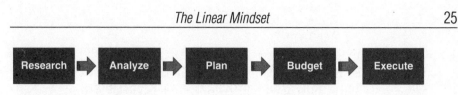

Figure 2.1   The Typical Corporate Planning Process

carefully, and then execute. Managers engage in an extensive process of analysis before taking action. They spend months and even years preparing their detailed plans. Throughout the planning process, managers build detailed forecasts based on past history as well as a number of critical assumptions. Once managers perfect their plans and budgets, they begin to implement. Much to their surprise and amazement, reality does not unfold according to their plans and projections in most circumstances. As Mike Tyson once remarked, "Everyone has a plan until they get punched in the mouth!"

Leonard Schlesinger, Charles Kiefer, and Paul Brown argue that this linear approach "works really well when the future can realistically be expected to be similar to the past."[6] However, novel situations and problems require us to act more like curious children than analytical MBA graduates. We certainly must do our homework, collect data, and formulate a strategy. Before too long, though, we must take actions, reflect, and learn, and then adapt accordingly (see Figure 2.2). We cannot expect to develop the perfect plan while sitting in our office amidst a pile of research reports. Instead, the creative problem-solving process involves going out into the world and engaging in a healthy dose of trial and error. We must learn by doing.

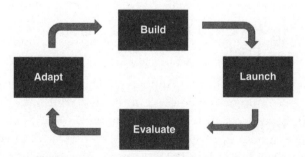

Figure 2.2   The Iterative Problem-Solving Process

For many corporate executives, though, this type of messy, non-linear process proves maddening. They crave straight lines, rather than a series of starts and stops, digressions, and feedback loops. A traditional planning or problem-solving process provides them a series of clear steps to follow. The steps often unfold as follows: frame the problem, develop alternatives, conduct analysis, select a course of action, and then implement the decision. This type of linear process offers managers the comforting feeling that they can forecast the future with precision, adhere to a set schedule, and control the flow of future events. Corporate strategy expert Russell Ackoff offers a humorous take on the way most executives solve problems and make decisions: "Most corporate planning is like a ritual rain dance: It has no effect on the weather that follows, but it makes those who engage in it feel they are in control."[7]

As da Vinci's work demonstrates, the creative process often comes fraught with tension, ambiguity, and a fair amount of dead-ends. Peter Carruthers of Los Alamos National Laboratory once said, "There's a special tension to people who are constantly in the position of making new knowledge. You're always out of equilibrium."[8] That feeling bothered him at first, but later he concluded that the most interesting learning often involved a healthy dose of discomfort. Da Vinci worked on many interesting things, and he certainly seems to have experienced that "out of equilibrium" feeling. This mode of thinking and working simply doesn't fit in many organizations today. Most executives prefer to hire specialists with deep yet narrow expertise, rather than curious polymaths. They desire control, certainty, and order. That preference for linear thinking blocks creativity in their organizations.

A Sanding Block, Deodorant Stick, and Butter Dish

Jim Yurchenco graduated from Antioch College in the early 1970s. He aspired to be a sculptor and enrolled in the master of fine arts program at Stanford. Several years later, he found himself working on designing the computer mouse for Apple's Lisa computer. How did that happen? As a sculptor, Yurchenco loved to hang out in the workshop where students in the Stanford product design program tackled various projects.

There he met a young engineer named David Kelley, who graduated from Carnegie Mellon in 1973 and went to work at Boeing. In one of his first assignments there, Kelley spent months designing the lavatory signs for the 747 jumbo jet.

After graduating from Stanford, Kelley partnered with another friend from school, Dean Hovey, to launch a product design firm. Kelley reached out to Yurchenco, who joined the firm as one of its earliest employees. Around this time, Apple became one of the young design firm's first clients. Steve Jobs needed a computer mouse for his Lisa computer. He had seen a mouse developed at Xerox's Palo Alto Research Center. However, Yurchenco remembers, "It was obviously way too complicated for what Jobs needed, which was a really low-cost, easily manufacturable, reproducible product for consumers."[9] According to some estimates, the Xerox mouse would cost $400 to manufacture. The Hovey-Kelley design firm set out to create a mouse for Lisa. As you might imagine, Steve Jobs proved to be a demanding client.

Yurchenco focused on the inner workings of the mouse, while colleagues tackled other elements of the design. The scrappy young firm did not have an abundance of financial resources. Computer-aided design software, 3D printers, and other sophisticated tools did not exist yet. Yurchenco recalls that they were always creating fast and crude prototypes.[10] A colleague, Jim Sachs, remembers cutting off the knob of a stick shift from a BMW to craft an early prototype. Kelley describes another early version of the mouse:

> I found one of these sanding blocks. You ever see these sanding blocks, where it's a big hunk of rubber and you put the sandpaper in it? I took one of those and chopped it up, and then put little divots in it like a golf ball, and painted two little eyes like a mouse, and that was the first mouse case. Apple rejected it completely.[11]

According to Yurchenco, they once purchased a roll-on deodorant stick and a butter dish from a local Walgreen's pharmacy. They needed the ball from the deodorant stick for one prototype, and the butter dish became the cover of the mouse. With each iteration, the designers learned what worked and what didn't. They gathered feedback from users and garnered Apple's reaction. The Lisa computer did not prove to be a success, but the Macintosh model later utilized the same design. The core elements of the Hovey-Kelley mouse design continued to be employed by computer makers for many years. Meanwhile, David Kelley began a decades-long relationship with Jobs and the design team at Apple.

Design Thinking Flourishes

In the early 1990s, Kelley, Bill Moggridge, and Mike Nuttall merged their design firms together to form IDEO.[12] Several years later, ABC's *Nightline* program chronicled the firm's efforts to design a new super-market shopping cart. The program brought a flood of attention to the company, its unique culture, and the creative process employed to design new products. Viewers learned that an eclectic team came together on each project. The shopping cart team featured a psychologist, linguist, biologist, and marketing expert working alongside engineers and designers—a *Renaissance team*, if you will. As Kelley noted:

> The point is that we're not actually experts at any given area . . . . We're kind of experts on the process of how you design stuff. So we don't care if you give us a toothbrush, a toothpaste tube, a tractor, a space shuttle, you know, a chair. It's all the same to us.[13]

Over time, many companies became very interested in learning more about the process employed by IDEO and other industrial designers, such as Maya, Frog, and Adaptive Path, to develop innovative products and services. As Kelley explained, "It's one thing to be able to do a product once in a while but if you can build a culture and a process where you routinely come up with great ideas, that's what the companies really want."[14] Organizations wanted to learn these methods and techniques. They became intrigued by the notion that teams could apply a replicable, systematic process to solve many types of complex problems, not just to design new products. Here was a roadmap that everyone could learn and apply. You didn't have to rely on the notion of hiring people who were born with special creative powers.

IDEO and other firms began to refer to the "design thinking process" to describe this creative problem-solving methodology, and they preached its applicability well beyond the field of industrial design. Tim Brown, Kelley's successor at IDEO, defines design thinking as "a human-centered approach to innovation that draws from the designer's toolkit to integrate the needs of people, the possibilities of technology, and the requirements for business success."[15] When Kelley stepped down as IDEO's CEO, he founded the d.school at Stanford University. The school's mission became to educate people on how to use the design thinking process to develop creative solutions to complex problems in a wide range of fields including health care, education, and business. Many other institutions began teaching design thinking, and corporations proved eager to learn this approach.

Each organization provided its own twist on the design thinking process, but these approaches tend to have a great deal in common. At Stanford, they describe the process as unfolding in five stages (refer to Figure 2.3). At the start, you must empathize deeply with the user. How are people behaving? What do they think and feel? Designers employ a variety of methods to develop empathy. Observation of users in their natural environment plays a central role. Using the ethnographic methods of anthropologists, they seek to understand people's behaviors, particularly those that are surprising or unexpected. They search for workarounds and adaptations (i.e. circumstances in which users have come up with makeshift solutions to address their frustrations or unmet needs). Design thinkers also learn by studying analogous experiences. Designers Rikke Dam and Teo Siang offer an example: "The highly stressful and time-sensitive procedure of operating on a patient in the emergency room of a hospital might be analogous to the process of refueling and replacing the tires of a race car in a pit stop."[16]

In the second stage, design thinkers try to make sense of the data that they have collected in the field. They synthesize what they have learned and search for patterns. They try to determine why people behave as they do. Ultimately, design thinkers strive to define a problem statement, or point of view, about user frustrations, needs, and desires. In stage three, design thinkers engage in idea generation. Using brainstorming and other techniques, they strive to generate a wide range of options for

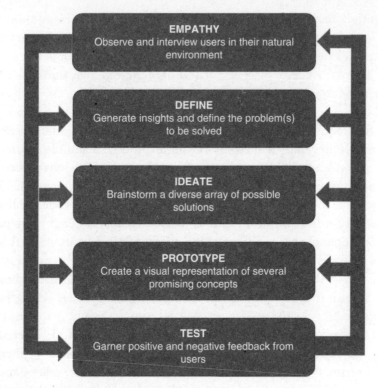

Figure 2.3   The Design Thinking Process

Source: Depiction of the process as characterized by the Stanford d.school[17]

addressing the design challenge that they have defined. During this stage, design thinkers adhere to Linus Pauling's philosophy: "The best way to have a good idea is to have a lot of ideas."

Prototyping takes place in the next phase of the process. Design thinkers do not select their best idea simply through extensive deliberation and analysis. They develop a series of low-fidelity prototypes—rough, inexpensive mock-ups that can be created quickly. Prototypes might come in many forms, including storyboards, role-play exercises, videos, and physical models—anything with which a user can interact in a meaningful way. In the final stage, design thinkers test their solutions by allowing users to provide feedback about multiple prototypes. They stage simple experiments to test hypotheses and explore unanswered questions. The learning that takes place during this testing mode informs further prototyping, and it guides further efforts to develop empathy, define user needs, and ideate.

Da Vinci: An Early Design Thinker

Notice the striking similarities between design thinking and the creative process employed by da Vinci in the 15th and 16th centuries. Before he began to invent and create, da Vinci engaged his marvelous powers of observation to understand people, systems, and natural phenomena. For instance, he looked carefully to see how blood flowed through the heart, light shined upon objects, birds flew through the sky, and body parts shifted as people moved. He documented these observations methodically, filling many notebooks with his sketches and comments. da Vinci did not rely only on what past experts said and wrote; he looked carefully himself to see how people, animals, or systems actually behaved and functioned.

Da Vinci employed analogous reasoning as he tried to understand how the world worked and developed his ideas. He looked for themes and principles that applied across disciplines. For instance, he drew analogies between how blood flowed through the human body and how water flowed on earth. Analogies can be dangerous at times, of course, because we can focus on the similarities between two domains and ignore crucial differences. Da Vinci understood the power and limitations of analogies. As he studied rivers and oceans, he recognized the flaws of his analogy to blood flow in the human body and adjusted his theories accordingly. Still, da Vinci's use of analogous reasoning enabled him to connect seemingly disparate concepts from different domains, and thereby make important new discoveries.

Much like today's design thinkers, da Vinci conducted experiments and developed many prototypes. Ian Hutchings, Professor of Engineering at the University of Cambridge, has argued that da Vinci conducted some of the earliest studies of the laws of friction, and that experiments played a key role in his investigative process. Hutchings has found evidence of these experiments in his exhaustive review of da Vinci's notebooks.[18] Similarly, da Vinci constructed rough prototypes throughout the process of painting, sculpting, and inventing. For instance, he once designed a mechanical knight, a 15th-century robot, if you will. Some experts believe he constructed and displayed it to visitors in Ludovico Sforza's court in Milan. Imagine the astonishment of the audience and the feedback he must have received![19]

Design thinking, then, does not represent an entirely new phenomenon. The origins of this way of creative thinking stretch as far back as the Italian Renaissance. Today's design thinkers learn by doing, much as Leonardo did centuries ago. They observe with a keen eye and "build to think" through sketches, storyboards, and mock-ups. They iterate often, learning and adapting quickly along the way. They are nonlinear thinkers, just as Leonardo was, and they are comfortable with ambiguity. Unfortunately, design thinkers experience something akin to organ rejection when entering many large organizations today.

Rough Sailing in Corporate Waters

Kaaren Hanson became one of the early advocates of design thinking at Intuit, the maker of personal finance, small business, and tax software. She worked closely with Intuit CEO, Brad Smith, and they helped turn Intuit into a "design-driven company" focused on creating delightful user experiences. The journey began at a company leadership conference in 2007. Smith asked executives to bring products to the meeting that they absolutely loved. People arrived with a range of items including a wine opener, sippy cup, and backpack.[20] Hanson says that this activity helped everyone understand the awesome feeling that they wanted Intuit's customers to experience. The transformation did not go smoothly though. Hanson points out that they made little progress during that first year.[21]

Slowly, though, Hanson and Smith began to instill design thinking throughout the organization. They appointed key people as Innovation Catalysts responsible for teaching and coaching others on design thinking methods and techniques.[22] They integrated design thinking into their leadership development efforts. Intuit developed methodologies such as *blueprinting*, a term the company uses to describe how cross-functional teams map and understand the user experience in great detail. Blueprinting involves mapping both the customer's journey (what Intuit calls the front-stage experience), and the organization's actions and processes behind the scenes to serve that customer (the backstage experience). Intuit designer Erick Flowers explains that blueprinting enables visualization of the end-to-end process of serving a customer.[23] Using these blueprints and other methods, Intuit discovered the ways to enhance the consumer's experience a great deal—to create *a-ha!* moments. Revenue grew by more than 70 percent over the past decade, and net income doubled. Smith credits the success, in large part, to the impact that design thinking has had throughout the company.

At IBM, Phil Gilbert has led a design revolution since he came onboard in 2010. Gilbert drew from many sources and created a proprietary design-thinking methodology for the company. He has hired an army of designers and opened studios around the world. With CEO Ginni Rometty's strong support, Gilbert also has invested to teach design thinking to a broad swath of IBM's employees. Gilbert believes that design thinking will help the company reverse a lengthy revenue decline associated with the deterioration of the company's traditional hardware businesses.[24]

Many other large companies have embraced design thinking in search of more rapid revenue growth and higher customer satisfaction. A.G. Lafley brought design thinking to Proctor and Gamble. The smash hit product Swiffer emerged from that effort. Gannett appointed Laura Ramos to use design thinking in an effort to reinvent the newspaper business. Russ Fleming, Michelle Proctor, and others developed FedEx's approach to design thinking. Patrick Douglas leads the IDEAS team at Target, leveraging human-centered design to enhance the store experience and cope with the e-commerce revolution.[25] Companies such as Pepsi and Philips Electronics have appointed chief design officers to serve on the top management team.

For many firms, though, design thinking has not led to the payoffs they anticipated. Executives invest large sums in training programs. Employees attend workshops, and they find the concepts and techniques fun and stimulating. Companies build idea labs and innovation hubs to support creative collaboration. They purchase mobile whiteboards and loads of Post-It notes. Unfortunately, the energy soon dissipates. Many ideas emerge from various workshops, but few of them stick. Prototypes never become finished products. Executives begin to bemoan the lack of return on investment. The pressure to produce results quickly leads to general disenchantment with design thinking. People begin to wonder whether it's just another management fad. Designer Tim Malbon, co-founder of Made by Many, offers a humorous, yet insightful take on the unrealized potential of many design thinking initiatives:

> Design thinking all too often delivers a wonderful day or two off from the realities of boring old business-as-usual. You'll always remember the day you were let out of your work cubicle (aka veal-fattening pen) to "play" with Sharpies and Post-Its along with colleagues on Brainstorm Island. You'll meet colleagues on a Design Thinking safari whom you may not even have realized before actually have first names. Working together—probably in teams—you'll be empowered to ideate the heck out of some really tough business problems, instead of just using boring old Excel and Word to count beans and execute your tiny piece of "The Plan." Afterwards, you'll probably tell loved ones at home about it, and you'll think back to it wistfully for years to come as you sit safely back inside your pen.... However, in terms of lasting, impactful, commercial innovation, the Design Thinking scorecard doesn't look so healthy. In reality, when you return from a trip to Brainstorm Island you probably won't have done any real innovation—at least, not the sort that's going to transform the fortunes of your business.[26]

Why has design thinking encountered roadblocks in many organizations? Many experts blame the culture of large companies. They are too bureaucratic and risk averse. Executives do not tolerate the failures that inevitably occur. These concerns certainly ring true; the problem runs much deeper, though. Linear thinking permeates most enterprises, whether in the private sector, the nonprofit world, or the government. Design thinking, in contrast, is a fundamentally non-linear process. Implementation does not follow formal planning in this approach. You build to think and iterate repeatedly, much as Leonardo did in all his creative endeavors. Adopting this creative approach requires a mindset shift.

Many companies have failed to make the shift from the traditional planning mindset to a learning-by-doing approach. Strategy formulation and implementation remain largely disconnected from one another. Firms continue to engage in annual strategic-planning rituals, pretending that they can predict the future from the confines of the corner office. Even worse, they have treated design thinking as just another linear process that they can deploy. Step two always follows step one. They march through the phases robotically, as if they have discovered a magic formula for innovation. Dam and Siang explain the flaw in this approach:

> The five stages are not always sequential—they do not have to follow any specific order and they can often occur in parallel and be repeated iteratively. As such, the stages should be understood as different modes that contribute to a project, rather than sequential steps.[27]

Trying to turn any creative process—design thinking or otherwise—into a highly structured, linear system turns out to be a colossal mistake. No one believes that a great artist can create *Last Supper* through a programmatic, paint-by-numbers approach. Why do business leaders believe that the iPhone, Amazon Echo, Ember mug, or Dyson vacuum should be any different?

We Hate to Iterate

Why can't organizations shed linear thinking and shift to a learning-by-doing mindset? It turns out that most of us don't enjoy iterating frequently, nor are we particularly good at it. Why? We become invested in a particular solution quite easily, and we do not handle feedback effectively. We know that learning and adaptation will enhance the quality of our solutions, but we fall in love with our initial ideas. We build mock-ups to validate concepts rather than to learn. Many people might say that they desire feedback, but most really crave praise and unconditional love. Nonlinear processes simply cause us far too much discomfort.

Imagine that you have invested a significant amount of time, money, and energy into a particular idea. The concept emerged after extensive user research, brainstorming, and storyboarding. What happens if users raise serious questions about its viability or feasibility during testing? We should let bygones be bygones, abandoning concepts that do not meet user needs. However, many organizations plow ahead, pouring good money and effort after bad. They rationalize or dismiss the concerns that users raise, and they convince themselves that just a little more investment will turn things around. Managers continue down the same path, perhaps with a few slight adjustments, rather than considering entirely different options.

Psychologists describe this phenomenon as the sunk-cost trap. We escalate our commitment to failing courses of action if we have made substantial prior investments in a particular initiative. Why? Self-justification, aimed at resolving cognitive dissonance, represents one possible cause of the sunk-cost effect. We experience dissonance when we develop an original idea, but then encounter failure or negative feedback. We cannot reconcile the results with our belief in the soundness and creativity of our initial idea. The further commitment of time, money, and effort provides an opportunity to resolve our dissonance. By plowing ahead, individuals can sustain the belief that this additional effort may turn around the situation and validate the original concept (as well as bolster the positive image we maintain of ourselves).[28]

Entrepreneurs speak often of the need to pivot when their startup encounters obstacles and challenges. They adhere to the lean-startup philosophy, which preaches the value of launching minimum viable products and iterating quickly based on customer reactions. Startups refer to legendary pivot stories such as the emergence of Twitter. The concept arose at a startup called Odeo that built a podcasting platform. In 2005, Apple announced its entry into the podcasting business. Given the strength of the iPod's installed base, the Odeo team knew that they were in trouble. The startup shifted gears, and it built the social media platform that became Twitter instead. In a little more than two years, the platform had 10 million users.[29] For every story such as this one, though, we can find dozens of startups that failed to pivot. Entrepreneurs fell in love with their initial idea and resisted change; or, they tinkered around the edges rather than considering bold new options when market success did not materialize.

Our inability to learn effectively from our mistakes also impedes the iterative process. Most people examine others' failures quite differently than their own mistakes. When others stumble, we look inside of them. We blame the failure on the poor choices they made and their lack of knowledge or expertise. We impugn their motives at times, or we conclude that certain personality flaws impaired their decision making. When we fail, we often do not look in the mirror. Instead, we blame environmental conditions or circumstances. We explain to ourselves that uncontrollable external factors led to the failure. Psychologists describe this phenomenon as the fundamental attribution error. By systematically blaming external factors for our failures, and ignoring internal causes, we fail to recognize opportunities for learning and improvement.[30]

When individuals do receive critical feedback, they can become quite defensive and stop listening. People tend to look for information that will confirm what they already believe, and they avoid or dismiss data that contradict their pre-existing views. Worse than that, we distance ourselves over time from people who don't tell us what we wish to hear. Paul Green, Francesca Gino, and Brad Staats have conducted some interesting research on this type of behavior. These scholars studied four years' worth of employee performance data from over 300 full-time workers at one particular company. At this organization, managers did not conduct annual performance reviews. Instead, people engaged in self-evaluation and they reviewed their peers. The researchers examined these data, as well as information about each worker's network within the organization. What did they find? Individuals tended to eliminate colleagues from their network if these co-workers provided negative feedback. If they could not exclude the person, they compensated by bringing others who would be more affirming into their social circle. In short, individuals surrounded themselves with people who told them what they wanted to hear. They paid a price for this behavior. The researchers found that employee performance suffered considerably when workers disassociated themselves from colleagues offering critical feedback.[31] Table 2.1 shows the types of thinking that prevent an effective iterative process.

Table 2.1   Impediments to an Effective Iterative Process

Problem	Definition	Consequence
Sunk-Cost Trap	Our unwillingness to cut our losses, and our tendency to throw good money and effort after bad.	We stick to, or perhaps tweak, our initial concept, rather than abandoning the idea and pivoting to a different concept.
Fundamental Attribution Error	Our tendency to blame others' failures on internal causes, while attributing our own failures to external circumstances.	We do not learn effectively from mistakes. We rationalize poor feedback from users in ways that prevent good adaptation.
Confirmation-Seeking Behavior	Our tendency to stop engaging with and listening to people if they provide negative feedback to us.	We wall ourselves off from the type of critical feedback that could be used to adapt and revise successfully.

Spaghetti and Marshmallows

Years ago, Peter Skillman, Director of Design at Skype, created a simple design exercise called the Marshmallow Challenge. Teams must construct a tall freestanding tower using 20 strands of spaghetti, a small amount of masking tape and string, and a single marshmallow. They have 18 minutes to complete the exercise. The marshmallow must remain intact, and it must rest atop the structure.[32]

Designer Tom Wujec loved the exercise and began conducting it with teams around the world. He has tested it with different types of audiences. As you may imagine, architects and engineers excel at the design challenge. Interestingly, recent graduates of business school struggle mightily. Many of their structures buckle and collapse. The average height of their structures falls far below average. Since these business school graduates perceive the marshmallow as lightweight, they underestimate the strength of the structure required to support it.[33]

One group of people tends to outperform the business school graduates by a wide margin. Elementary school children excel at the

Marshmallow Challenge! Why? Wujec observes that the business school alumni tend to proceed in a linear fashion. They assess the task and the materials provided, and they formulate a plan for constructing the tower. These teams engage in a fair amount of debate and deliberation as to the most effective design. After finalizing a design, they begin to build. As the deadline approaches, they finally pick up the marshmallow and place it atop the structure. A few teams celebrate, but many watch in horror as the spaghetti snaps and the tower crumbles.[34]

Wujec argues that business people learn in school to find the one best solution through analysis and deliberation. They master techniques for systematic evaluation of data. They fall in love with the quantitative rigor provided by methods such as net present value analysis, multiple regression, and the like. Business school alumni conclude that mistakes are costly, and that disciplined preparation can avoid those costly errors. In short, many business school students learn to think in a linear manner. However, in many ambiguous, novel situations, "Enlightened trial and error succeeds over the planning of the lone genius," as Peter Skillman has argued.[35]

Many business school graduates struggle because they approach the Marshmallow Challenge with a *how/best* mindset. As University of San Diego Professor Jennifer Mueller explains, people with this mindset tend to focus on searching for and identifying the one best solution to a problem. They place a premium on accuracy and correctness in decision making. They often question how a novel idea will be implemented and what obstacles might prevent successful adoption, rather than focusing on why it might just work. The how/best mindset represents linear thinking—one path, and one path only, must lead to an optimal solution. Creativity suffers when people adopt a how/best mindset. In contrast, Mueller finds that creativity flourishes when individuals embrace a *why/potential* mindset. People with this attitude tend to approach problems with the belief that multiple viable solutions exist. They do not look for the one perfect path to success. These individuals ask positive questions about many alternatives: Why might this work? Why might this be a good idea? Why might we want to pursue this approach?[36]

How does the children's approach to the Marshmallow Challenge differ from the men and women in the business world? The kids tend to pick up the marshmallow fairly early in the process. After all, a marshmallow proves pretty enticing to a six year old! The children play with the marshmallow and the other materials. They do not spend a lot of time trying to draft the perfect design on paper. Instead, the children start building right away. They engage in trial and error, testing out what type of structure might hold the marshmallow. In short, the children adopt a why/potential mindset and embrace the iterative process, but the business school graduates do not. The children naturally gravitate toward prototyping and experimentation. After all, they don't learn how to walk by reading a book or listening to a lecture. They let go of mom's hand, take a step or two, and stumble and fall. They pick themselves up, and they try again. Just as they discovered how to walk, they learn how to build a strong tower. They learn by doing.

Endnotes

1. My uncle, Arcangelo Ranaudo, took me to this museum in Anchiano, Italy, several times over the years. The museum is located roughly 45 kilometers from the city of Florence.
2. This section draws from Walter Isaacson's excellent biography of da Vinci. See Walter Isaacson, *Leonardo Da Vinci* (New York: Simon and Schuster, 2017).
3. Paula Rae Duncan, *Michelangelo and Leonardo: The Frescoes for the Palazzo Vecchio,* Graduate Thesis, University of Montana, 2004.
4. Isaacson, *Leonardo Da Vinci*, 82.
5. Ibid., 87.
6. Leonard Schlesinger and Charles Kiefer, *Just Start: Take Action, Embrace Uncertainty, Create the Future* (Boston: Harvard Business Review Press, 2012), 9.
7. Russell Ackoff, "On the Use of Models in Corporate Planning," *Strategic Management Journal,* 2(4), 1981, 359.
8. William J. Broad, "The Creative Mind: Tracing the Skeins of Matter," *New York Times Magazine,* May 6, 1984. (www.nytimes.com/1984/05/06/magazine/the-creative-mind-tracing-the-skeins-of-matter.html?pagewanted=all, accessed November 6, 2017).
9. Kyle VanHemert, "The Engineer of the Original Apple Mouse Talks About His Remarkable Career," *Wired,* August 18, 2014 (www.wired.com/2014/08/the-engineer-of-the-original-apple-mouse-talks-about-his-remarkable-career/, accessed November 6, 2017).

10. Alex Pang and Wendy Marinaccio, "Interview with Jim Yurchenco and Rickson Sun," *Making the Macintosh: Technology and Culture in Silicon Valley* website, June 10, 2000 (web.stanford.edu/dept/SUL/sites/mac/primary/interviews/ideo/index.html, accessed November 6, 2017).

11. David Kelley qtd. in Alex Pang and Wendy Marinaccio, "The Apple Mouse," *Making the Macintosh: Technology and Culture in Silicon Valley* website, September 12, 2000 (web.stanford.edu/dept/SUL/sites/mac/primary/interviews/kelley/mouse.html, accessed November 6, 2017).

12. I learned about the history of IDEO through the sources cited in this chapter, as well as through attendance at a series of workshops at IDEO in March 2016. The workshops included a presentation by Tim Brown, IDEO's CEO. I also had the opportunity to interview IDEO's Alan Ratliff about the history and culture of IDEO.

13. *Nightline*, "The Deep Dive," ABC, February 9, 1999.

14. Ibid.

15. IDEO U, "Design Thinking," 2018 (www.ideou.com/pages/design-thinking, accessed November 7, 2017).

16. Interaction Design Foundation, "Stage 1 in the Design Thinking Process: Empathise with Your Users," 2017 (www.interaction-design.org/literature/article/stage-1-in-the-design-thinking-process-empathise-with-your-users, accessed November 9, 2017).

17. Thomas Both and Dave Baggeroeor, "(Archival Resource) Design Thinking Bootcamp Bootleg," Hasso Plattner Institute of Design at Stanford University, n.d. (dschool.stanford.edu/resources/the-bootcamp-bootleg, accessed March 23, 2018).

18. Ian Hutchings, "Leonardo da Vinci: The First Systematic Study of Friction," University of Cambridge Department of Engineering website, May 5, 2016 (www.eng.cam.ac.uk/news/leonardo-da-vinci-first-systematic-study-friction, accessed November 10, 2017).

19. Isaacson, *Leonardo da Vinci*. See also Zaria Gorvett, "Leonardo da Vinci's Lessons in Design Genius," BBC Future, July 28, 2016 (www.bbc.com/future/story/20160727-leon, accessed November 3, 2017).

20. Brad Smith, "Intuit's CEO on Building a Design-Driven Company," *Harvard Business Review*. January-February 2015.

21. "Intuit: How Design Drove Its Turnaround," *Bloomberg Businessweek*, March 20, 2014 (www.bloomberg.com/news/articles/2014-03-20/intuit-how-design-drove-its-turnaround, accessed November 8, 2017).

22. Roger Martin, "The Innovation Catalysts," *Harvard Business Review*, June 2011.

23. I heard Flowers describe Intuit's blueprinting process at Google's Design Sprint Conference, November 2017.

24. Personal interview with IBM's Phil Gilbert, November 2014. Personal interview with IBM's Charles Hill, November 2015. Personal tour of IBM Design Studio in New York, October 2015.

25. Personal interviews with Laura Ramos, Michelle Proctor, and Patrick Douglas.

26. Tim Malbon, "The Problem with Design Thinking," Made by Many, March 3, 2016 (www.madebymany.com/stories/the-problem-with-design-thinking, accessed November 10, 2017).

27. Rikke Dam and Teo Siang, "5 Stages in the Design Thinking Process," Interaction Design Foundation, 2017 (www.interaction-design.org/literature/article/5-stages-in-the-design-thinking-process, accessed November 9, 2017).

28. For more on why we exhibit sunk-cost bias, see Barry Staw and Jerry Ross, "Understanding Behavior in Escalation Situations," *Science,* 246, 1989, 216–220; Hal Arkes and Catherine Blumer, "The Psychology of Sunk Cost," *Organizational Behavior and Human Decision Processes*, 35, 1985, 124–140; Joel Brockner, "The Escalation of Commitment to a Failing Course of Action," *Academy of Management Review*, 17(1), 1992, 39–61. For more information on how to avoid the sunk-cost trap, see Michael Roberto, "Cutting Your Losses: How to Avoid the Sunk Cost Trap," *Ivey Business Journal,* November/December 2009. Personal interviews with high-altitude mountaineers David Breashears and Ed Viesturs have informed my understanding of why human beings experience sunk-cost bias and how to overcome it.

29. Adam Penenberg, "An Insider's History of How a Podcasting Startup Pivoted to Become Twitter," *Fast Company,* August 9, 2012 (www.fastcompany.com/1837848/insiders-history-how-podcasting-startup-pivoted-become-twitter, accessed November 15, 2017).

30. Lee Ross and Richard Nisbett, *The Person and the Situation: Perspectives of Social Psychology* (London: Pinter & Martin, 2011).

31. Paul Green, Jr., Francesca Gino, and Bradley Staats, "Shopping for Confirmation: How Disconfirming Feedback Shapes Social Networks," *Harvard Business School Working Paper 18-028*, September 2017.

32. https://www.tomwujec.com/marshmallowchallenge, accessed November 17, 2017.

33. Tom Wujec, "Build a Tower, Build a Team," TED Talk, February 2010 (www.ted.com/talks/tom_wujec_build_a_tower, accessed November 17, 2017).

34. Personally, I have conducted the Marshmallow Challenge with several thousand students as part of Bryant University's IDEA program. I've also used the exercise in many executive education settings. My results confirm Wujec's findings. Most business students and managers fail to engage in prototyping during this exercise.

35. *Nightline*, 1999.

36. Jennifer Mueller, *Creative Change: Why We Resist It . . . and How We Can Embrace It* (Boston: Houghton Mifflin Harcourt, 2017).

CHAPTER 3

The Benchmarking Mindset

The one who follows the crowd will usually get no further than the crowd.
The one who walks alone is likely to find themselves in places no one
has ever been before.

—Albert Einstein, physicist

As a teenager, Mark Burnett left his family's East London home and joined the British Army. He became an elite paratrooper and served in the Falkland Islands conflict. After leaving the army, Burnett chose to embark on another adventure. He moved to Los Angeles with $600 in his pocket. He worked as a nanny for several families when he first arrived in the United States. Burnett also sold used clothing at several different beaches in southern California.

One day he became inspired by an article describing the Raid Gauloises, an adventure/endurance race created by Gerard Fusil. The former paratrooper felt that television viewers might be mesmerized by "people on the edge of death, racing for a prize."[1] Burnett teamed up with Brian Terkelsen, a former Wall Street banker with expertise in the media and entertainment industry. They created the *Eco-Challenge*, with five-person teams traversing 300 miles of mountains, canyons, cliffs, and rivers in southeastern Utah. Burnett billed the challenge with the motto, "This Little Race Eats Ironmen for Breakfast."[2] Over the course of eight grueling days, the team members hiked, ran, and bicycled their way through the wilderness. They traveled by raft, canoe, and horseback. Only 21 of the 50 registered teams completed the race.

Burnett and Terkelsen convinced MTV to air the event, and it garnered quite a bit of attention including a feature on the Dateline NBC program. ESPN aired the next *Eco-Challenge*, set in Maine, as part of the X-Games. Discovery Channel and USA Network televised the show in subsequent years.[3]

Burnett landed his next idea for a television program when he met British producer Charlie Parsons at a party. Parsons wanted to make a show in which contestants competed for $1 million while stranded on a deserted island. Burnett purchased the rights to produce and air the show in the United States, but he struggled to find a network that would televise the program. He pitched the idea to Fox, ABC, CBS, NBC, Discovery Channel, and USA Network. They all rejected him. Finally, he persuaded Les Moonves, head of CBS Television, to give his show a chance. Moonves would not commit to televise this rather bizarre-sounding program during the fall prime-time schedule though. Instead, he decided to air the show during the summer, not exactly an attractive landing spot on the broadcast network calendar.[4]

Survivor aired for the first time on May 31, 2000, hosted by Jeff Probst. Sixteen castaways competed on the remote Malaysian island of Pulau Tiga. The show became a surprising smash hit over a 13-week run. *Time* and *Newsweek* ran cover stories about America's voyeuristic fascination with the show. Rhode Island native Richard Hatch, known for walking around the island naked much of the time, eventually won the $1 million prize. A stunning 51 million people tuned into the August 23 finale, far more than the number of people who watched the World Series or the season finale of the wildly popular NBC hit *Friends*.[5]

A wave of new reality television programs hit the airwaves in the years that followed. CBS premiered *Big Brother*, Fox launched *Temptation Island*, and ABC debuted *The Mole* in 2001. The *Washington Post* estimates that over 300 reality television programs aired over the next 15 years. Some shows, such as *American Idol* and *The Bachelor*, became long-running hits. Quite a few others (e.g., *I Want to Be a Hilton*, *Celebrity Boxing*) came and went quickly without causing much of a stir or attracting many viewers.[6]

Similar bouts of herd behavior have occurred throughout Hollywood history. For example, in the early to mid–1960s, westerns outnumbered police dramas by a wide margin on the three major television networks. In fact, police dramas represented only 1 percent of the shows aired in 1964. By the end of the decade, a few promising new police shows debuted, including *Hawaii Five-O* and *Ironside.* When those two programs garnered strong ratings, Hollywood started producing an avalanche of police dramas (see Figure 3.1). In 1975, a whopping 28 percent of programs airing in prime time were cop shows, including *Police Woman, Starsky and Hutch,* and *Kojak.* The western genre had disappeared completely by the end of the decade.[7]

After more than a few major flops, the police drama waned in popularity in the late 1970s. CBS premiered a steamy prime-time soap opera named *Dallas,* starring Larry Hagman and Linda Gray. The program chronicled the rather sleazy behavior of oil tycoon J.R. Ewing and his absurdly dysfunctional family. The show rocketed to the top of the ratings, becoming the number-one show in America by its third season. Copycats soon followed, including *Knots Landing, Falcon Crest,*

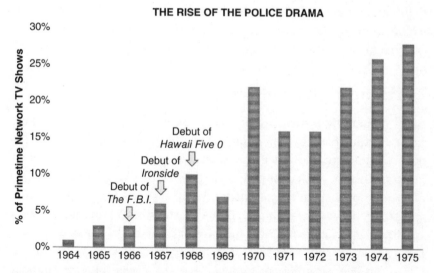

Figure 3.1 Herd Behavior in Network Television Programming

Source: Data drawn from Robert E. Kennedy, "Strategy Fads and Competitive Convergence."[8]

Dynasty, and *Flamingo Road.*[9] The police/crime drama roared back in the mid-1980s though, particularly due to the emergence of a number of shows featuring private detectives. Again, a few breakout hits, such as *Magnum, P.I.* and *Simon and Simon,* spawned a surge of imitators.

Business economist Robert Kennedy has conducted a systematic analysis of herd behavior in Hollywood. He examined all prime-time broadcast network programs aired between 1961 and 1989. His dataset included 975 television shows from that era. The average program lasted a little more than two seasons. The data demonstrated that the popularity of various genres shifted considerably over time. Kennedy used a sophisticated statistical model to determine the prevalence of imitation versus differentiation strategies as the networks developed and launched new shows. In other words, were the networks predominantly following trends or trying to stand out with distinctive programming? Not surprisingly, he concluded that ABC, CBS, and NBC tended to mimic one another a great deal. That would make perfect sense if emulating hit shows proved fruitful. However, Kennedy discovered that distinctive, original shows consistently outperformed trend followers. Specifically, imitation strategies led to lower ratings and shorter program longevity than differentiation. Why would the networks behave in an apparently irrational manner for three decades? Why didn't they learn that creating original ideas pays off better than simply mimicking your rivals?[10]

Learning from the Best

Many organizations spend a great deal of time and money benchmarking against leading firms in their industries and beyond. They measure themselves against the elite, identify performance gaps, and search for best practices that they can adopt. Why do they do it? Clearly, imitation pays off handsomely at times, particularly when it's used to enhance quality and productivity. One such benchmarking success story unfolded several years ago at Cincinnati Children's Hospital Medical Center. Dr. Uma Kotagal, an executive leader and senior fellow, earned a $1.9 million "Pursuing Perfection" grant from the Robert Wood Johnson Foundation. She used the proceeds to launch two major improvement initiatives, one of which focused on the quality of care for cystic fibrosis patients.[11]

Kotagal and her colleagues gathered data from the Cystic Fibrosis Foundation about patient outcomes at medical centers around the country. The results concerned her a great deal. The Cincinnati Children's Hospital had dropped to the 20th percentile in the nation on a critical measure, patient lung functioning. Nutritional status, as measured by the patient's body mass index, also lagged behind the majority of other medical centers. The results surprised many of Kotagal's colleagues, who had believed for years that their hospital provided some of the very best care in the country.

Kotagal and her team asked the Cystic Fibrosis Foundation to provide them the names of the five hospitals generating the best patient outcomes for the disease. Then they visited and/or spoke by phone with medical professionals at each hospital, and they identified the practices and techniques that led to high performance. The team instituted major changes as a result of this learning initiative. For instance, Kotagal and her colleagues focused on improving airway clearance for patients, a critical daily process by which cystic fibrosis sufferers remove mucus from their lungs. Seven years later, the Cincinnati Children's Hospital Medical Center ranked in the 95th percentile in the nation in terms of patient lung functioning. The benchmarking effort had a discernible positive impact on the lives of many patients.

The Benchmarking Curse

Benchmarking sometimes may lead to important operational or quality improvements, but all too often, it causes firms to adopt copycat strategies and business models. As management expert Gary Hamel says, "Strategies converge because success recipes get lavishly imitated... Aiding and abetting strategy convergence is an ever-growing army of eager young consultants transferring best practice from leaders to laggards."[12] Rather than leveraging employee creativity to forge a distinctive path, firms choose to emulate their rivals' product offerings and competitive positioning. As head-to-head competition intensifies, total industry profits tend to decline.

How and why does benchmarking and competitor analysis lead to strategy convergence? Harvard marketing professor Youngme Moon describes how competitors typically react after conducting competitive analysis. Suppose that Firm #1 finds itself lagging behind the competition on product attribute X. Its competitor, Firm #2, excels in that area, but it does not perform well with regard to product attribute Y. What happens after both firms study each other closely? Each organization focuses on improving its most glaring weakness or vulnerability. Firm #1 seeks to catch up to its rival with regard to attribute X, and Firm #2 strives to bolster attribute Y. Both companies seek to become well-rounded, meaning they want to perform reasonably well across all attributes that are important to customers in that particular industry. Soon, though, the two companies look alike. She explains that, "Well-meaning efforts to monitor your competitive position . . . can turn into a cattle prod for homogenization."[13]

Moon explains that firms should instead recognize that differentiation comes from becoming more lopsided rather than well rounded. You establish a distinctive competitive position by amplifying your strengths, rather than engaging in knee-jerk efforts to imitate your competitors. To be great, you must reject doing some things at which others excel. Take, for example, the outdoor power equipment industry. Stihl ranks as the top-selling chainsaw brand in the world. While many rivals have lowered costs by outsourcing component manufacturing, the family-owned Austrian company continues to engineer and produce saw chains and guide bars in-house to maintain the highest quality standards. When others built strong relationships with big box stores such as Home Depot and Lowe's, Stihl chose not to sell their products through these channels. Stihl only distributes its equipment through a network of 45,000 dealers around the world. Why? The company aims to provide exceptional service before and after the customer's purchase. Their dealers assemble products for consumers and they offer expert training.[14] Several years ago, Stihl even went so far as to run advertisements boasting that consumers could not find their products at big box stores. One tagline read: "You see, we won't sell you a Stihl in a box, not even a big one."[15] Through this strategy, Stihl has established a unique position in the industry and enjoyed a great deal of success.

Building and sustaining competitive advantage means making trade-offs, explicitly choosing what not to do. You have to zig when others zag. Steve Jobs once said, "I'm actually as proud of many of the things we haven't done as the things we have done."[16] Great organizations recognize that you cannot be world class in all areas simultaneously. Investing in excellence in one area inevitably means making sacrifices along other dimensions. Ferrari would find it difficult to build the fastest, sleekest, most luxurious sports car in the world while also ranking at the top of the industry on child safety.[17] Unfortunately, too many firms try to be all things to all people, and thus, we witness herd-like behavior across many industries.

People often ask legendary British entrepreneur Richard Branson to share his secrets for success. How does one become abundantly wealthy? Branson once joked, "If you want to be a millionaire, start with a billion dollars and launch a new airline."[18] Indeed, the airline industry has been one of the least profitable industries on earth for more than a century. Think about all the bankruptcies in the past few decades: Continental (1983), Eastern (1989), Pan Am (1991), Northwest (2005), and many, many more. TWA filed for bankruptcy three times between 1992 and 2001![19] In Warren Buffett's 2007 letter to the shareholders of Berkshire Hathaway, he described the airline industry as a gruesome business. He joked, "Here a durable competitive advantage has proven elusive ever since the days of the Wright Brothers. Indeed, if a farsighted capitalist had been present at Kitty Hawk, he would have done his successors a huge favor by shooting Orville down."[20]

Many reasons exist for this unfortunate predicament, but strategy convergence certainly plays a key role. Jost Daft and Sascha Albers studied the 26 firms in the European airline industry between 2004 and 2012.[21] Not surprisingly, they discovered that the firms became more similar to one another during this era. For instance, Daft and Albers documented how the airlines' fleets and route networks became more homogenous. One competitor stood out from the rest, becoming more distinctive over time. That company, Ireland's low-cost player Ryanair, generated some of the highest profits in the European airline industry during this period. In the United States, Southwest Airlines resisted the temptation to imitate more established rivals for years, and it generated strong profits year after year while many competitors went bankrupt.

Imitation efforts not only fail because they create more pernicious and intense head-to-head competition within an industry, while stifling creativity and imagination. Firms do not falter simply because they copy each other, but because they copy badly! In many instances, companies do not truly understand what makes rivals successful. They study a competitor closely and identify a few elements of the business model that they deem worthy of emulation. They mistakenly credit those factors with a rival's success. Companies often do not recognize that competitive advantage rests on building an integrated system of activities.[22] Success does not depend upon a silver bullet, but on the well-tuned alignment of brand positioning, operational activities, corporate culture, and so on. For great companies, the whole is truly worth more than the sum of the parts, yet firms often imitate a few isolated elements of an industry leader's business model.

The Scandinavian retailer Ikea has enjoyed extraordinary success in the furniture industry for many years. Have competitors tried to discern the sources of the company's competitive advantage? You bet. Have they been successful emulating the company? Not so much. Former Ikea president Anders Dahlvig explains the many factors that contribute to high customer satisfaction and strong financial results. These include product design, distribution, store layout and atmosphere, and the like. Then, he notes, "Many competitors could try to copy one or two of these things. The difficulty is when you try to create the totality of what we have."[23] You cannot imitate a few parts of the model and achieve similar outcomes, yet firms often do just that when they benchmark industry leaders.

Managerial career concerns provide one final possible explanation for copycat behavior in many industries. People benchmark and imitate market leaders because that conservative path tends to provide job security.[24] Imagine that the leaders in your industry have all chosen to embark upon a similar strategy. What will happen if you choose a radically different course of action? Typically, embarking on such a path represents a high-risk–high-reward proposition. You might soar to new heights and surpass the competition, or you could crash and burn. As a manager, what happens if you follow the crowd? Your performance will hover around the industry average. You may not be rewarded handsomely, becoming rich beyond your wildest dreams. In all likelihood, though, matching your peers' performance will provide you job security. Suppose that you choose to swim into unchartered waters. You might produce exceptional performance. However, if you fail and perform well below the industry average, you may lose your job. Concerns about job security, therefore, cause many managers to pursue the conservative, follow-the-crowd approach.

Dave Grohl's Inspiration

Pablo Picasso once said, "Good artists borrow. Great artists steal." Studying others' amazing successes need not diminish our creativity. Sometimes, our aspirations grow when we observe what others have achieved, and we can find inspiration by examining the great work of predecessors and contemporaries in our field. Many breakthroughs in science, business, and the arts emerge from combining past concepts in novel, unexpected ways. Analysis of exemplars need not lead to copycat behavior.

Musician Dave Grohl illustrates how creativity can flow from the inspiration provided by past greats. At age 10, Grohl began to teach himself how to play guitar. His mother purchased the Beatles' greatest hits albums (commonly called *The Red Album* and *The Blue Album*) for him, as well as the sheet music for the iconic bands' entire catalog. Grohl recalls:

> That's really where I learned to play guitar. I would put an album on, find the page with the song, try to play along according to this simple music sheet, almost like I was in a band in my bedroom, trying to follow along with these other players. I'd have to remember an arrangement, and changes and tempo and melody. So those two albums were my music teacher when I was young.[25]

Grohl loved the song "Rocky Raccoon" and performed it at age 12 in front of a large audience of high-school students. He also became fascinated with "Paperback Writer," a number-one song written in 1966 by John Lennon and Paul McCartney. He says, "It had that nasty groove to it. I thought it was great that they could look like such gentlemen and sound like such bad asses."[26] As Grohl learned to play the drums, he would emulate Ringo Starr and other drummers, such as John Bonham of Led Zeppelin.

Grohl's fascination did not end in his childhood years. As he became an adult, he continued to look to the Beatles for inspiration. In 1990, Grohl joined the path-breaking group Nirvana as its drummer. Nirvana emerged from the grunge scene in Seattle in the late 1980s. Lead singer Kurt Cobain wrote the songs for that group. Interestingly, he too looked to the Beatles for inspiration. He had learned to sing "Hey Jude" when he was a preschooler. Cobain recalled, "My aunts would give me Beatles records, so for the most part [I listened to] the Beatles [as a child], and if I was lucky, I'd be able to buy a single."[27] For Nirvana's first album, he wrote a song titled *About a Girl*, after listening to a Beatles album for hours.

While Nirvana enjoyed remarkable success, Grohl wrote and recorded songs privately. He recalls keeping the songs to himself, because he did not want to interfere with Cobain's creative process. Meanwhile, he continued to enjoy playing Beatles' music. In 1994, he played drums for a group that recorded cover songs of numerous Beatles tunes for the soundtrack to the movie *Backbeat*.[28] When Cobain died tragically in 1994, Nirvana disbanded. Grohl began recording the songs that he had written over the years. He sang all the vocals, and he played many instruments. Soon he formed his own band, Foo Fighters, drawing inspiration from Cobain as well as past groups such as the Beatles, ABBA, Queen, and many others. Grohl loved to learn from and pay homage to past rock greats. Still, Foo Fighters developed their own distinctive style. Foo Fighters went on to record many hits and earn numerous Grammy Awards.

Throughout his career, Grohl recognized that imitation simply would not work. He describes trying to emulate the "Ringo roll," a particular technique used by the Beatles drummer. Grohl laughs when he recalls his failed attempts: "Whenever I try to do it, I look like I'm putting out a fire with a jacket!"[29] Grohl explains that you must be authentic. He says, "Being a musician should be, you know, not to really force it so much . . . just kind of let it happen. That's the idea, I think, is to just be yourself."[30]

Grohl's fascination with the Beatles came full circle in 2007. Grohl decided to record a cover of McCartney's *Band on the Run* at London's famous Abbey Road studios. He met McCartney for the first time when the Beatles legend came by the recording session. Grohl felt as though he had died and gone to heaven. The two musicians soon became friends. Later that year, Grohl played drums while McCartney performed "I Saw Her Standing There" and "Back In The U.S.S.R." at a concert in Liverpool.[31] Several years later, McCartney joined the surviving members of Nirvana and recorded a new song for a Hurricane Sandy benefit concert.[32] Imagine being a fly on the wall during those jam sessions! The group's new tune, "Cut Me Some Slack," earned a Grammy Award as Best Rock Song of the year. The relationship continued in 2017, when McCartney played the drums on a song for Foo Fighters' ninth album, *Concrete and Gold*.[33] Grohl's journey demonstrates that you can learn from, become inspired by, and even collaborate with idols without becoming a bland imitator. In the end, he paved his own path, creating a distinct sound and style that attracted a host of fans.

Fixation and Water Jars

Unfortunately, imitation supersedes inspiration for many of us. Humans often experience fixation when trying to solve a problem. That is, we become attached to a specific mental set, a way of thinking about a problem based on solutions that have worked in the past. In 1903, Orville and Wilbur Wright's plane did not have flapping wings. Others had tried this approach previously because they allowed a mental set to constrain their problem-solving process. Observing that birds flew by flapping their wings, a series of inventors including E.P. Frost and Gustave Trouvé attempted to mimic this approach for aircraft.[34] Mental sets can facilitate problem solving at times, but becoming fixated on an inappropriate solution from past experience can inhibit creativity.

Abraham Luchins illustrated this type of fixation through his famous water-jar experiments. He asked people to solve a series of problems such as the following. You have three jars (A, B, and C). A can hold 21 units of water, B can hold 127 units, and C can hold 3 units. You must find a way to measure out precisely 100 units of water using these three jars. What is the solution? The simple equation [B − A − 2C] provides the answer. You begin by filling Jar B with water. Then, you pour as much of that water as you can into Jar A. At this point, you have 106 units of water left in Jar B. Now you fill Jar C with water twice, and that will leave precisely 100 units in Jar A.

In the experimental condition, Luchins provided subjects with a series of practice problems for which an algorithm always provided the solution. The control group did not receive the practice problems. All participants later received a water-jar problem that could not be solved using the algorithm. The participants in the experimental condition struggled, as they attempted unsuccessfully to use the previously useful equation to solve this novel problem. Luchins also gave both groups several water-jar scenarios for which an easier method could be used to arrive at a solution. The control group, who did not receive any practice problems, tended to use this easier algorithm while the others often did not. Those individuals, conditioned by the practice problems, again relied on the tried-and-true formula that had worked for them in the past, even though a simpler method could have been utilized. Amazingly, when Luchins gave participants the simple message "Don't be blind" it increased the likelihood that people would adopt the easier technique![35]

Does fixation inhibit creativity in product development? Imagine that you are a designer or engineer. If you were trying to develop a new product, you might begin by examining a range of existing designs. You would want to know what worked and what didn't in the past. David Jansson and Steven Smith conducted a series of experiments in which they tested whether students and professional engineers might be susceptible to design fixation. In one study, they asked a group of Texas A&M mechanical engineering students to design a bicycle rack for an automobile. Jansson and Smith provided one half of the students with a diagram of a bike rack. The sketch showed a rack mounted on top of a vehicle's roof, with suction cups as part of the system holding it in place. Railings helped keep the tires of each bike secure. The scholars pointed out one weakness in that sketch, namely that it would be challenging to mount the bike in the middle of the roof rack. The control group simply went about its task without seeing a prior design. The scholars asked all participants to generate as many conceptual designs as possible in one hour. Interestingly, the two groups generated an equal number of designs. However, those individuals exposed to a previous design demonstrated clear signs of fixation. Their sketches showed more top-mount designs, more inclusion of suction cups, and more tire railings than the control group's work.[36]

Surely, design fixation only holds when we benchmark reasonably good designs that others have created, right? If engineers examine prior designs with many flaws, perhaps susceptibility to fixation diminishes. Jansson and Smith set out to test this theory in a follow-up experiment. They asked mechanical engineering students to design a disposable, spill-proof coffee cup. The instructors provided one half of the students a sketch and pointed out many flaws, including some issues with regard to the use of a straw and mouthpiece in the design. They also directed the students not to use a straw or mouthpiece in their sketches. The control group simply began their work without looking at the flawed design that someone else had created. Once again, both groups came up with roughly the same number of designs in one hour. However, the individuals that viewed the prior work showed clear signs of fixation. Elements of the old design crept into their work in various ways. Some students even employed a straw or mouthpiece in their design, despite the explicit instruction not to do so!

Jansson and Smith measured the creativity of the students' work along two dimensions, originality and flexibility (see Figure 3.2). A student scored high in originality if they produced a large number of distinctive designs. Individuals scored high in flexibility if they generated a diverse set of approaches to creating the perfect cup. As you might expect, people who viewed a sample sketch exhibited lower originality and flexibility. Even studying a highly flawed exemplar led to fixation and diminished creativity.

What if we provided people examples of past work, but told people not to imitate that design? Smith and his colleagues conducted a subsequent study to examine this question.[38] They asked people to design new toys. In this case, they did not provide designs and point out imperfections. Instead, they simply showed people some examples of prior designs. They instructed some students as follows: "We have found that examples like those you examined restrict people's creativity. Try NOT to restrict your ideas. When the drawing task begins, please generate

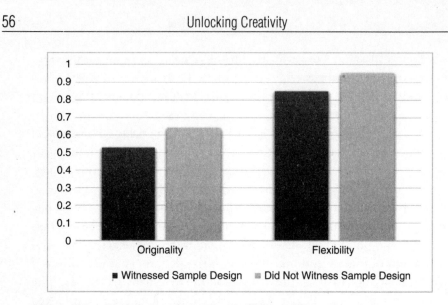

Figure 3.2   Design Fixation: The Spill-Proof Cup Experiment
Source: Data drawn from Jansson and Smith, "Design Fixation"[37]

ideas as different as possible from the examples given."[39] Surprisingly, the researchers discovered that this instruction did not reduce the tendency to conform to the sample designs initially shown to the individuals. People find it difficult to erase the benchmark from their minds and resist imitation, even when clearly instructed to do so. Talk about being stuck in your head!

Look to the Outside!

How can organizations protect themselves against fixation? For starters, we can seek to learn from people outside our industry, rather than strictly benchmarking our direct rivals. Searching for inspiration from other fields often proves to be very stimulating, particularly if we can find analogous experiences or products to examine. Reebok did just that when it created the famous Pump technology for its athletic shoes.

In the late 1980s, Nike enjoyed a remarkable run of success thanks to the Air Jordan line of sneakers. These shoes not only featured an endorsement from the greatest basketball player on earth, but also incorporated a balloon filled with compressed gas that provided extra comfort and support in the heel of the sneaker. Reebok's revenues and profits stagnated, as people purchased several million Air Jordans in their first year on the market. Reebok's CEO assigned engineer Paul Litchfield to design a sneaker that could compete with Nike's innovative new product line.

Litchfield sought assistance from Design Continuum, an innovation and design agency based in Massachusetts. The team that came together to work for Reebok included designers with experience in the healthcare field. Certain medical products proved quite analogous to the new type of heel the team hoped to build. One designer had experience designing an inflatable splint. What if you could employ a "splint in a shoe" design of some sort? Could it be used to enhance ankle support in a basketball sneaker, and thereby reduce the risk of injury? Another designer had worked on creating better intravenous fluid bags for hospitals. That product inspired the team to consider building an inflatable bladder into the heel of the sneaker. How could an athlete inflate this bladder, thereby customizing the shoe in a meaningful way? The team looked to others with experience designing diagnostic instruments for medical professionals. These instruments often contained pumps, valves, and tubing. A blood pressure monitor did the trick. It inspired the team to devise a solution for *pumping up* the heel of the sneaker.

The sneaker hit the market at the end of 1989, and it generated $500 million in sales that first year. Boston Celtics star Dee Brown pumped up his Reebok sneakers before launching himself in the air during the NBA's All-Star Weekend Dunk Competition in 1991. When Brown took top honors, Reebok Pump sales soared to incredible new heights. Litchfield's team did not imitate the Air Jordan; that would have been a hopeless endeavor. Instead, they created a distinctive and exciting shoe with a little help from their friends in healthcare.[40]

Marion Poetz, Nikolaus Franke, and Martin Schreier conducted an interesting study to evaluate the impact of soliciting inspiration from analogous fields.[41] (See Figure 3.3.) They sought ideas for how to encourage safety-gear compliance from skaters, carpenters, and roofers. They challenged these people to offer solutions for enhancing the comfort of safety belts, respirator masks, and knee pads so that people would use them more often. Individuals tended to offer more novel concepts when thinking of solutions for fields other than their own. In

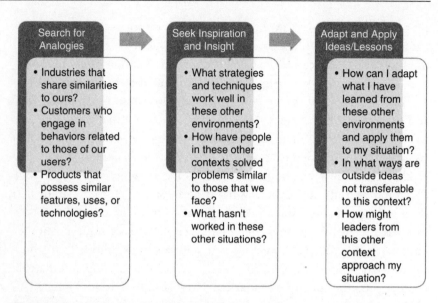

Figure 3.3   Learning from Analogous Contexts

fact, novelty increased as they moved farther from their area of expertise and specialization. For instance, they asked everyone to come up with ideas for better respirator masks used by carpenters. As it turns out, the roofers came up with more novel ideas than the carpenters, but the inline-skaters offered the most original solutions of them all!

Define Substitutes Broadly

What kind of fitness center offers pizza, bagels, and Tootsie Rolls to its members? Planet Fitness, of course! The New Hampshire–based chain has achieved its success by pursuing a rather unconventional approach. *Men's Health* magazine once published an article with the headline: "Planet Fitness Is Not a Gym: And It's Stupid To Keep Pretending It Is."[42] Blog posts and online discussions often feature fitness enthusiasts ridiculing the chain. Indeed, the company offers limited free weights, choosing to focus on cardio equipment such as treadmills and exercise bicycles. Company spokesman McCall Gosselin explains, "Our clubs don't have equipment like squat racks and Olympic benches. Our dumbbells only go up to 80 pounds."[43] Serious fitness enthusiasts have mocked the company for its lack of proper weight-training equipment.

Planet Fitness enjoys telling the world that it does not want hard-core bodybuilders in its gyms. In fact, employees can sound a "lunk alarm" if someone behaves in a way that might intimidate the average person trying to grab a quick workout. Television advertisements boast that the gyms are "judgement-free zones." One famous commercial features a bodybuilder receiving a tour at one of the company's gyms. When asked what type of workout he enjoys, the man with bulging muscles and *very* short, *very* tight pants keeps repeating, "I lift things up and put them down." Eventually, the employee tricks the bodybuilder into stepping outside and locks the door behind him. The advertisement ends with the words "not his planet . . . yours" appearing on the screen.[44]

Why would Planet Fitness go out of its way to signal that certain people are not welcome at its fitness centers? The answer is actually quite simple. The fitness chain wants everyone to know the type of people that it intends to serve and the type of environment it wishes to create. Founder Marc Grondahl explains, "We are going after the 85 percent of the population that isn't hard-core fit. We want people to feel comfortable and accepted here no matter what their workout level."[45] The company does not view traditional fitness centers as its main competitors. Planet Fitness CEO Chris Rondeau explains:

> The first main competitors are honestly Chili's and Uno's and the movie theaters. Other brands look at working out as a hobby, and I think personally that working out is a chore, and I believe most of America thinks of it the same way, they know they have to [but] they'd rather go to Chili's and have a beer and have some chips and salsa, but you know, you have to, you don't want to, so you kind of wince your way throughout. And I think most of Americans think this way.[46]

Why would Planet Fitness pursue such an unusual strategy? The firm recognized that a conventional strategy would yield low profits. Few barriers to entry exist in this fragmented industry, and gyms compete intensely on price. Customers often demonstrate more loyalty to their personal trainers than to their fitness centers. If a beloved trainer departs to a gym down the street, exercise enthusiasts often follow along. Consumers have abundant choices amongst relatively undifferentiated gyms. Moreover, consumers can choose from a plentiful array of substitutes to going to the gym. Individuals can work out at home, go on a diet, exercise outdoors, or even undergo weight-loss surgery. Of course, some people may choose not to work out at all. As a result of this intense competition, many fitness centers have failed over the years. One-time industry leader Bally Total Fitness entered bankruptcy on two occasions. Many other fitness centers, particularly independent gyms, have faced similar struggles.

Planet Fitness chose a different path, and the company has managed to survive in a very challenging environment for several decades. Founded in 1992, the company now has over 10 million members in more than 1,500 locations across the nation. Will it survive and thrive for years to come? Many people have doubts. Some analysts have raised questions about the dizzying pace of growth and the rapid expansion through franchising. Perhaps the unconventional strategy will falter amidst intense competition and overly ambitious growth objectives. In all probability, though, simply copying the industry leaders would have led to dismal results. Give credit where credit is due. The company tried something new when so many players adhered to the conventional wisdom and chose to proceed down the well-trodden path, with modest profits at best in most circumstances. Planet Fitness instead chose a riskier path in hopes of defying the odds in a very challenging industry environment. They might not sustain their success for years to come, but they have made a valiant effort in a brutally tough industry.

Rather than simply benchmarking direct rivals, companies need to think broadly about the full range of substitutes against which they compete (see Table 3.1). American Airlines competes against Skype and WebEx, not just Delta and United. Because of these technologies, we can conduct business with colleagues in Japan and Italy without boarding a plane to those countries. Progressive and Amica compete against Uber, as these firms will sell fewer insurance policies if more people choose not to purchase an automobile.

Table 3.1   Examples of Substitution Threats

Product/Service	Potential Threat of Substitution	Rationale
Chewing Gum	Amazon	If people shop online more frequently, fewer opportunities will exist for impulse purchases of chewing gum in retail stores.
Luggage Carts	Airbnb	If people opt to use Airbnb rather than stay at hotels while traveling, demand for luggage carts may decrease.
Fine China	Expedia	If engaged couples choose travel experiences over kitchenware on their wedding registry, then demand for fine china may fall.
Auto Insurance	Uber, Lyft	If some people choose not to purchase cars, opting to use these services instead, then the insurance market may shrink.
Airlines	Skype, WebEx, Polycom	If businesspeople adopt these communication technologies, they may not need to travel as frequently for meetings.

Planet Fitness did not fixate on other gyms. Moreover, they did not simply think about other exercise or weight-loss options as substitutes for visiting a fitness center. They recognized that people might simply choose not to work out. They could watch a movie or go out to dinner in their free time. Planet Fitness did not focus on wooing customers who had chosen to sign up with their competitors. They decided to target people who typically did not work out at a gym at all. They asked, "Why don't some people visit fitness centers, and what alternative activities do they pursue in their free time?" By defining and evaluating substitutes broadly, Planet Fitness built a highly distinctive strategy.

Taking the Leap

Veering away from the herd requires courage. When you choose not to imitate, you often make bold bets that may lead to failure. When you embark on a creative new endeavor, you must be willing to withstand ample skepticism and even ridicule, as has been the case with Planet Fitness. In many instances, people do not recognize the merits of a distinctive, creative strategy at first. Some organizations find it easier to fall back in line, blending into the crowd, when they encounter resistance to new ideas. Others persist. They learn and adapt when obstacles emerge, and they press forward.

Survivor rocketed to great success in a matter of a few weeks. Most creative new endeavors do not become overnight sensations. Many television shows face cancellation unless they achieve such early success. In 1995, Donald Bellisario produced a unique new show for NBC. The program, *JAG*, focused on the exploits of Harmon Rabb, a former fighter pilot, and other members of the U.S. Navy's Judge Advocate General's office. At that time, the networks produced very few shows focused on the military. NBC canceled the program after just one season due to poor ratings and lackluster reviews from critics. Bellisario did not quit. He retooled the show and CBS picked up the program in January 1997.[47] *JAG* lasted nine more seasons, and it routinely found itself among the top 25 most-watched prime-time programs each week. In 2003, the spin-off *NCIS* premiered on CBS. The program, starring Mark Harmon, focused on the members of the Naval Criminal Investigative Service. Bellisario hit the jackpot. *NCIS* has aired for 15 seasons and has been the number-one drama on prime-time television for years.[48] Bellisario chose a different path. While many other producers jumped on the reality-show bandwagon, he took a chance on a different type of program. His experience demonstrates that not following the herd can be a rocky and risky path, but it might just be incredibly rewarding.

Endnotes

1. Bill Carter, "Survival of the Pushiest," *The New York Times*, January 28, 2001 (www.nytimes.com/2001/01/28/magazine/survival-of-the-pushiest.html, accessed November 27, 2017).

2. Mark Burnett, *Jump In! Even if You Don't Know How to Swim* (New York: Ballantine Books, 2006).

3. Personal interview with Brian Terkelsen, March 2017.

4. Bill Carter, *Desperate Networks* (New York: Crown Publishing, 2007).

5. Steve Johnson, "'Survivor' Finale Posts Ratings Even Larger Than Show's Hype," *Chicago Tribune*, August 25, 2000 (articles.chicagotribune.com/2000-08-25/news/0008250272_1_survivor-finale-nielsen-ratings-cbs-spokesman-chris-ender, accessed November 27, 2017).

6. Emily Yahr, Caitlin Moore, and Emily Chow, "How We Went from 'Survivor' to More Than 300 Reality Shows: A Complete Guide," *Washington Post*, May 29, 2015 (www.washingtonpost.com/graphics/entertainment/ reality-tv-shows/, accessed November 27, 2017).

7. Robert E. Kennedy, "Strategy Fads and Competitive Convergence: An Empirical Test for Herd Behavior in Prime-Time Television Programming," *The Journal of Industrial Economics,* 50(1), 2002, 57–84.

8. Kennedy, "Strategy Fads and Competitive Convergence," 2002.

9. PBS, "Primetime Soaps," *Pioneers of Television* website, 2014 (www.pbs.org/wnet/pioneers-of-television/pioneering-programs/primetime-soaps/, accessed November 28, 2017).

10. Kennedy, "Strategy Fads and Competitive Convergence."

11. Anita Tucker and Amy Edmondson, "Cincinnati Children's Hospital Medical Center," Harvard Business School Case Study 9-609-109, April 25, 2011.

12. Gary Hamel, *Leading the Revolution* (Boston: Harvard Business School Press, 2000), 49.

13. Youngme Moon, *Different: Escaping the Competitive Herd* (New York: Crown Business, 2010), 37.

14. The Stihl Group, "The Stihl Group: Leading the Way for More Than 90 Years," Stihl website, n.d. (www.stihl.com/about-stihl.aspx, accessed November 30, 2017).

15. Stihl, Advertisement in *Winnipeg Free Press*, June 4, 2008, 22 (newspaperarchive .com/winnipeg-free-press-jun-04-2008-p-22/, accessed November 30, 2017).

16. Betsy Morris, "Steve Jobs Speaks Out," *Fortune*, March 7, 2008 (archive.fortune .com/galleries/2008/fortune/0803/gallery.jobsqna.fortune/6.html, accessed November 30, 2017).

17. Moon, *Different.*

18. Henry Truc, "A Bittersweet Moment for Richard Branson as Virgin America Sells to Alaska Air Group," Equities.com, April 4, 2016 (www.equities.com/news/a-bittersweet-moment-for-richard-branson-as-virgin-america-sells-to-alaska-air-group, accessed December 1, 2017).

19. Airlines for America, "U.S. Airline Bankruptcies," Airlines.org, n.d. (airlines.org/dataset/u-s-bankruptcies-and-services-cessations/, accessed December 1, 2017).

20. Berkshire Hathaway, Annual Report, 2007.

21. Josh Daft and Sascha Albers, "An Empirical Analysis of Airline Business Model Convergence," *Journal of Air Transport Management*, 46, 2015, 3–11.

22. Michael Porter, "What Is Strategy?" *Harvard Business Review*, November–December 1996.

23. Ken Favaro and Art Kleiner, "The Thought Leader Interview: Cynthia Montgomery," *Strategy + Business* 70, Spring 2013 (www.strategy-business.com/article/00163?gko=f0c49, accessed December 1, 2017).

24. For more on how career concerns shape managerial decisions, see Jeffrey Zwiebel, "Corporate Conservatism and Relative Compensation," *Journal of Political Economy*, 103(1), 1995, 1–25.

25. Mojo Staff, "Born to Run," *Mojo* magazine, October 2017 (www.fooarchive.com/features/mojo2017.htm, accessed December 2, 2017).

26. Michael Heatley, *Dave Grohl: Nothing to Lose*, 4th edition (London: Titan Books, 2011), 10.

27. Sarah Anderson, "50 Things You Never Knew about Nirvana," *NME*, September 13, 2011 (www.nme.com/photos/50-things-you-never-knew-about-nirvana-1413101, accessed December 2, 2017).

28. Kory Grow and Jonathan Bernstein, "Dave Grohl's Guest List: 21 Amazing Musical Cameos," *Rolling Stone*, January 11, 2018 (www.rollingstone.com/music/lists/dave-grohls-guest-list-15-amazing-musical-cameos-20141119/the-backbeat-band-1994-w515308, accessed December 2, 2017).

29. Dave Grohl, Interview, Arclight Theater, Hollywood, CA, October 23, 2013 (www.youtube.com/watch?v=9dRG_PaZYjM, accessed December 2, 2017).

30. Ibid.

31. Shirley Halperin, "Paul McCartney and Dave Grohl: A Bromance Is Born Before the Grammys," *Rolling Stone*, February 8, 2009 (www.rollingstone.com/music/news/paul-mccartney-and-dave-grohl-a-bromance-is-born-before-the-grammys-20090208, accessed December 2, 2017).

32. Billboard Staff, "Paul McCartney, Nirvana Alums Rock 'Cut Me Some Slack' at Sandy Concert," Billboard.com December 13, 2012 (www.billboard.com/biz/articles/news/1483945/paul-mccartney-nirvana-alums-rock-cut-me-some-slack-at-sandy-concert, accessed December 2, 2017).

33. Ryan Reed, "Paul McCartney Plays Drums on Upcoming Foo Fighters LP," *Rolling Stone*, August 2, 2017 (www.rollingstone.com/music/news/paul-mccartney-plays-drums-on-upcoming-foo-fighters-lp-w495625, accessed December 2, 2017).

34. Wright Brothers Aeroplane Company, "A History of the Airplane," 2010 (www
.wright-brothers.org/History_Wing/History_of_the_Airplane/History_of_the
_Airplane_Intro/History_of_the_Airplane_Intro.htm, accessed December 4,
2017).

35. Abraham Luchins, "Mechanization in Problem Solving: The Effect of
Einstellung." *Psychological Monographs,* 54(6), 1942.

36. David Jansson and Steven Smith, "Design fixation," *Design Studies,* 12(1), 1991,
3–11.

37. Ibid.

38. Steven Smith, Thomas Ward, and Jay Schumacher, "Constraining effects of
examples in a creative generation task," *Memory & Cognition,* 21(6), 1993,
837–845.

39. Smith, Ward, and Schumacher, 843.

40. Jake Rossen, "Adjusted for Inflation: A History of the Reebok Pump," Mental
Floss, October 22, 2015 (mentalfloss.com/article/69922/adjusted-inflation-
history-reebok-pump, accessed December 6, 2017).

41. Marion Poetz, Nikolaus Franke, and Martin Schreier, "Sometimes the Best Ideas
Come from Outside Your Industry," *Harvard Business Review* (Digital Article),
November 21, 2014.

42. Lou Schuler, "Planet Fitness Is Not a Gym," *Men's Health,* January 11, 2014
(www.menshealth.com/fitness/planet-fitness-is-not-a-gym, accessed
December 7, 2017).

43. Schuler, 2014.

44. Planet Fitness, "Lift Things Up" commercial, 2011
(www.youtube.com/watch?v=q7gzmoqmL7g, accessed December 7, 2017).

45. Katie Morell, "Marc Grondahl of Planet Fitness: How a Lean Business Model
Became a Franchise Heavyweight," American Express Open Forum, June 13,
2013 (www.americanexpress.com/us/small-business/openforum/articles/mark-
grondahl-of-planet-fitness-how-a-lean-business-model-became-a-franchise-
heavyweight/, accessed December 7, 2017).

46. Mallory Schlossberg, "Planet Fitness's main competitors are Chili's and Uno's,
CEO says," *Business Insider,* August 17, 2016 (www.businessinsider.com/planet-
fitness-ceo-says-competition-isnt-gyms-2016-8, accessed December 7, 2017).

47. Paul Brownfield, "'JAG' Is Alive and Well at CBS," *Los Angeles Times,* May 4,
1998 (articles.latimes.com/1998/may/04/entertainment/ca-46102, accessed
December 8, 2017).

48. Sandra Gonzalez, "How 'NCIS' Became the World's Most-Watched Show,"
Mashable, September 22, 2015 (mashable.com/2015/09/22/ncis-worldwide-
viewers/#Pe7sbThnAZqww6, accessed December 8, 2017).

CHAPTER 4

The Prediction Mindset

The best way to predict the future is to invent it.
> —Alan Kay, computer scientist

Jim Cramer rants and raves on television each night on the popular CNBC show *Mad Money*. He throws chairs, wears costumes, and tapes Post-it notes to his forehead in a format that some have described as "more like pro wrestling than *Wall Street Week*."[1] Cramer does not do nuance. He takes bold stands and draws definitive conclusions. Hundreds of thousands of viewers tune in each weekday evening as the Harvard-educated former hedge-fund manager offers investment advice. It's certainly entertaining television. Is he a phenomenal prognosticator though? Wharton students Jonathan Hartley and Matthew Olson decided to find out. They examined the performance of Cramer's Action Alerts PLUS portfolio from 2001 to 2016. The investment fund includes many recommendations put forth on Cramer's hit television program. Lo and behold, this portfolio has underperformed the Standard and Poor's 500 Index over a fifteen-year period.[2] Give Cramer credit. You will learn something amidst all the histrionics on the show, as he does offer insight and knowledge based on decades of experience on Wall Street. Don't be mistaken though. You won't find predictions marked by uncanny accuracy, and you won't become wealthy beyond your wildest dreams.

The Jim Cramer phenomenon exists in many fields. Prognosticators command tremendous attention in our society. People look to accomplished economists to forecast future economic growth and unemployment levels. We listen to former superstar athletes and coaches picking the winners of next weekend's football games. At the end of each year, many magazines publish entire issues dedicated to predictions about the year ahead. People have an incredible desire to see the future and to find those who can help them achieve that objective. Amazingly, we continue to have great faith in expert forecasts, despite considerable evidence suggesting that we should look upon their predictions with a healthy dose of skepticism. Experts may know a great deal about the past, having built up an impressive amount of knowledge in a particular domain. They can perform remarkable feats of analysis. Forecasting is another matter altogether.

Nobody Knows Anything

Hollywood directors and producers wish that they could predict the future. Each year, they sift through many pitches and read tons of scripts. They examine all types of creative ideas from talented writers. Studio heads have to make some big bets, as production budgets have exploded in recent years. Many films cost more than $100 million to produce, and then studios must incur huge marketing expenses as well. Breaking even on a film can be a formidable challenge. The presence of a big-name actor or actress in a movie by no means guarantees success. As it turns out, forecasting box office receipts turns out to be just as difficult as beating the stock market, if not more so.

In 1997, Ben Affleck earned an Academy Award for Best Original Screenplay with his partner, Matt Damon, for the movie *Good Will Hunting*. The film cost roughly $10 million to produce and grossed over $225 million globally.[3] One year later, he co-starred with Bruce Willis in the box office smash, *Armageddon*, and he had a supporting role in the Oscar-winning film, *Shakespeare in Love*. Affleck became one of Hollywood's biggest stars, commanding more than $10 million per film. In 2002, *People* magazine named Affleck the sexiest man alive.[4] One year later, his romance with pop music superstar and actress Jennifer Lopez dominated tabloid headlines. The two decided to star in a movie together. It cost $74 million to produce, yet only generated $7 million in box office receipts. The completely forgettable *Gigli* became one of the biggest box office bombs in movie history.[5]

Many other award-winning actors, directors, and producers have suffered a similar fate. Academy Award winner Geena Davis once starred in a movie called *Cutthroat Island*. The movie cost more than $100 million to produce, yet it only grossed $10 million at the box office. Comedy legend Eddie Murphy played the leading role in a science fiction picture titled *The Adventures of Pluto Nash*. That film cost $120 million to produce, but only generated $7 million in ticket sales. The 2001 ensemble comedy, *Town and Country*, featured Warren Beatty, Goldie Hawn, and Diane Keaton. It cost $105 million to make, yet box office receipts totaled just $10 million.[6]

Author and screenwriter William Goldman once wrote that "nobody knows anything" in the movie business. Writing in 1983, he cited the remarkable success of *Raiders of the Lost Ark,* the fourth highest grossing film of all-time at that point. George Lucas wrote and Steven Spielberg directed the film, which starred Harrison Ford as archaeologist and adventurer Indiana Jones. The film grossed nearly $400 million at the box office, yet cost only $18 million to produce.[7] Goldman points out that every single major Hollywood studio rejected the film, except Paramount. He comments:

> Why did Paramount say yes? Because nobody knows anything. And why did all the other studios say no? Because nobody knows anything. And why did Universal, the mightiest studio of all, pass on *Star Wars*... because nobody, *nobody*—not now, not ever—knows the least goddamn thing about what is or what isn't going to work at the box office.[8]

Hollywood's experience over the years supports Goldman's contention. Each year, several big-budget movies starring leading actors and actresses bomb at the box office. At the same time, a few films become surprise hits. In 2017, the movie *Wonder* exceeded expectations by a wide margin when it topped $132 million in box office receipts.[9] Famous sleeper hits of the past include *Easy Rider, My Big Fat Greek Wedding,* and *The Blair Witch Project*—all low-budget films that attracted huge audiences in theaters. Former Columbia Pictures President David Picker might have been right when he told Goldman, "If I had said yes to all the projects I turned down, and no to all the ones I took, it would have worked out about the same."[10]

People in other fields have experienced similar frustrations with regard to predicting success. In the National Football League, the fate of a franchise often rests on the shoulders of its quarterback. Teams find it difficult to win consistently without a top-notch player at this crucial position. Coaches and general managers conduct in-depth evaluations of college players each year before the NFL draft. Each team must determine whom to select with their precious first-round draft pick. Over the years, though, many quarterbacks drafted in the first round have been major failures. Infamous busts include Johnny Manziel, JaMarcus Russell, and Ryan Leaf. In contrast, sixth-round draft pick Tom Brady has won more Super Bowls than any other quarterback in history. Since 2000, two thirds of the quarterbacks selected in the first round failed to become stars in the league (see Figure 4.1). Just 4 of these 48 quarterbacks have won Super Bowl championships.[11] Wharton

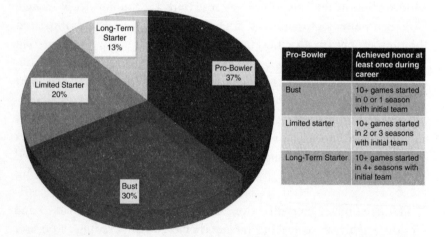

Figure 4.1   National Football League Quarterbacks Drafted in the First Round: 2000–2014

Source: Analysis conducted by author based on data from Pro Football Reference[13]

Professor Cade Massey has studied the NFL draft and he concludes that it's a major crapshoot. He says:

> Some teams have great years, other teams have bad years—and it matters. But those differences aren't persistent year to year, which tells me that they are chance driven. Something between 95 and 100 percent—I'm not exaggerating—of team differences in the draft is driven by chance.[12]

A Dart-Throwing Chimpanzee

Philip Tetlock has studied the human capacity for forecasting for many years. Starting in the 1980s, Tetlock analyzed the predictions of 284 experts, half of whom had doctorates in their fields, on issues of international politics and economics. These individuals worked at universities, think tanks, government agencies, and international organizations such as the World Bank. He collected data for 21 years, so that he could document what actually happened over time and compare it to these initial predictions. As it turns out, "The average expert did about as well as random guessing."[14] The running joke in the popular press became that a dart-throwing chimpanzee can predict the future about as well as any so-called expert. Tetlock's studies did not demonstrate that all experts are useless, as some in the popular press claimed, but he did demonstrate the limits to our predictive powers.

Years later, Tetlock launched the Good Judgment Project. He not only studied the accuracy of expert forecasts, but he also recruited over 2,000 people from all walks of life to make predictions about world events, economic trends, and the like. For instance, Tetlock and his colleagues asked people to answer questions such as, "Will Serbia be officially granted European Union candidacy by 31 December 2011?" and "Will there be a violent confrontation between China and one or more of its neighbors by March 2014 in the South China Sea?" He compared the results of these amateur forecasters with the work of intelligence analysts working for the federal government.[15]

Tetlock unearthed a very small subset of people who were amazingly good forecasters, better than most domain experts. For instance, Doug Lorch, a retired IBM computer programmer from California, made remarkably accurate predictions about international political and economic events. He had no particular expertise in international affairs, and Tetlock naturally did not provide him access to the type of classified information possessed by national intelligence professionals. Nonetheless, Lorch proved to be remarkably prescient. Tetlock coined the term "superforecasters" to describe those individuals who delivered consistently excellent predictions. These people weren't necessarily smarter than the experts; they did behave differently though. Tetlock discovered that *how people think* matters more than *what they already know.*

How did the best prognosticators behave differently than the vast majority of those who tried to predict the future? One superforecaster, Reed Roberts, argues that experts "often force new information into a pre-existing mental framework, or discard it if it seems to contradict their initial view."[16] In contrast, Tetlock's superforecasters tend to be open-minded, reflective, and intellectually curious. They acknowledge what they do not know. They gather information from a wide variety of sources and question the validity of each source. These individuals enjoy pondering a range of diverse views, and they update their conclusions as facts change. Superforecasters do not become wedded to any particular idea; they treat beliefs as testable hypotheses rather than hard truths. They don't necessarily have higher raw intelligence than the rest of us. They do exhibit a set of behaviors, mindsets, and habits that lead to higher levels of critical thinking and more accurate analysis.[17]

Sadly, many technical specialists, thought leaders, and executives appear to abandon these styles of thinking and reasoning as their reputations and accomplishments in a particular field grow. Consequently, accurate predictions remain elusive in many fields. In fact, Tetlock's research shows that widely recognized and frequently quoted experts tend to offer less reliable predictions than less well-known and cited specialists. The shocking political events of 2016 remind us of the challenges most prognosticators face. Few experts predicted the outcome of the Brexit vote in the United Kingdom, or the results of the presidential election in the United States. We live in a world with more data at our fingertips than ever before, yet surprises abound. Our obsession with prediction continues though. We keep hoping that people can see the future, no matter the dismal track record of past prophets.

The Need for Control

If many forecasters have such a poor track record, why do we continue to have such strong interest in predictions of all kinds? Put simply, human beings have a powerful desire for certainty. Psychologists Arie Kruglanski and Donna Webster describe this phenomenon as the need for cognitive closure.[18] Moreover, human beings have a powerful need to feel in control. Jennifer Whitson and Adam Galinsky write, "The desire to combat uncertainty and maintain control has long been considered a primary and fundamental motivating force in human life."[19]

How important is the sense of control? Studies show that if individuals feel in control of their lives, they tend to experience better physical and mental health. Nicholas Turiano and his colleagues examined this question using data from the national survey of Midlife in the United States (MIDUS). In the mid-1990s, over 7,000 randomly selected people ages 25 to 75 completed this survey. Turiano's research examined the factors that affected mortality rates, controlling for demographics and education levels. The researchers found that a sense of control reduced mortality risk, particularly for those with lower levels of educational attainment.[20]

Perhaps not surprisingly, human beings sometimes conclude that they have a sense of control in circumstances where they clearly do not. Ellen Langer has conducted a series of experiments to demonstrate what she calls the illusion of control, the feeling that we have a better chance at success in a particular situation than what objective probability would dictate. In one study, she showed participants a black wooden box with a glass top. Inside, subjects could see three intertwined paths etched in copper. A stylus placed on the paths completed a circuit and caused a buzzer to sound. Langer informed the participants that only one path would cause the buzzer to ring during each trial, and that she had programmed the machine to select that path randomly. She asked the participants to attempt to guess the path that would trigger the buzzer when the stylus was placed upon it.

Langer manipulated the extent to which subjects were involved in the game. Some participants placed the stylus on the path themselves. In other cases, Langer used the stylus at the direction of the participants. She also varied the extent to which the research subjects had a chance to practice with the machine. Some participants had an opportunity to become familiar with the machine and practice with the stylus for several minutes. Others had no such opportunity. Before the actual play began, Langer asked participants to indicate how confident they were about selecting a route that would ring the buzzer. Then she ensured that the first trial would be successful for all participants. Next she asked the subjects how they would perform against a champion chess player in five trials with this machine. Amazingly, those individuals who had used the stylus themselves and practiced beforehand exhibited much more confidence than those who had directed Langer to use the stylus and did not have a chance to practice. Introducing factors from situations where skill clearly does drive success, such as involvement and practice, into a game of chance led to irrational and unwarranted feelings of confidence. Hands-on involvement and prior practice had triggered an illusion of control. As a result, people think that they can predict something that they clearly cannot.[21]

What happens if individuals do not have a sense of control? They search for patterns even where they do not exist. They exhibit irrationality in drawing conclusions about cause and effect. For instance, a student might become convinced that he performs better on tests when using a particular green pen. Whitson and Galinsky write, "The need to be and feel in control is so strong that individuals will produce a pattern from noise to return the world to a predictable state."[22] The scholars tested this hypothesis through a series of experiments. In one study, they asked subjects to vividly recall a circumstance in which they either had a great deal of control or not. Then they presented the participants with three scenarios. Each one presented a rather dubious cause-and-effect relationship, one that might be deemed a superstition. For example, does knocking on wood translate into a better result at work? The individuals who had remembered a situation in which they lacked control tended to perceive a more powerful causal relationship than those who recalled being in control in a past situation. Put simply, we will draw spurious conclusions about what causes certain outcomes simply to create a sense of control when we lack that feeling!

Taken together, this series of studies demonstrates how powerful our need for certainty and control can be. We want desperately to believe that we can predict the future. If necessary, we will delude ourselves into thinking that we can forecast what will happen next. Many blame the corporate obsession with revenue and earnings forecasts on Wall Street's relentless demand for quarterly estimates. Perhaps, though, the obsession with prognostication runs much deeper than we thought. It appears to be deeply rooted in human nature.

Moving the Needle

Why does the obsession with prediction impede creativity in many organizations? To answer this question, we must understand how new projects get funded in many companies. Many executives set arbitrary sales targets regarding new products and services. They will only invest in a project if the result will move the needle in terms of top-line revenue. For many companies, that means they want new products to achieve at least $50 million in sales. If the innovation cannot deliver those types of results, many executives are not interested.

Unreasonable growth expectations often drive this concern about moving the needle with any new product or service. Imagine that you lead a company that generated $20 billion in revenue last year, ranking you among the 150 largest public companies in the United States in 2017. Setting a 10 percent growth target means aiming to generate $2 billion in new revenue next year, and thereby trying to double the size of the firm in roughly seven years. Achieving that type of growth organically can be very daunting. Thus, executives in large companies often express disinterest in niche products that may only achieve a few million dollars in revenue in the near term.

Growth and Expectations in the Magic Kingdom

In the early 1980s, Disney's earnings had declined for several consecutive years. The company spent considerable amounts of money completing the EPCOT theme park and launching the Disney Channel. Meanwhile, the quality of creative output at the film studio had been lackluster, and box office performance suffered as a result. Corporate raider Saul Steinberg launched a hostile takeover attempt in June 1984. Later that summer, Disney rebuffed Steinberg's attempt, as oil and gas investor Sid Bass took a large equity stake in the firm. Michael Eisner, former President of Paramount, became CEO of the floundering entertainment company.[23]

Eisner established a goal of growing revenues and earnings by 20 percent per year. In 1984, the company generated $1.66 billion in sales. Eisner updated and improved the theme parks, while raising prices considerably. He hired Jeffrey Katzenberg, who led a remarkable turnaround at Disney's film studio. From 1989 to 1994, Disney's animators produced a series of smash hits including *The Little Mermaid, Beauty and the Beast, Aladdin,* and *The Lion King.* Eisner also capitalized on the explosion of the home video market in the 1980s, releasing many classic Disney movies on videocassette. He opened Disney retail stores around the world, and he grew the company's cable television operations.

By 1994, Disney's revenues climbed past $10 billion, achieving Eisner's target of a 20 percent compound annual growth rate over 10 years. To this day, many people consider it to be one of the most remarkable turnarounds in recent history. 1994 proved to be a troubling year, though. Eisner's top lieutenant, Frank Wells, died tragically in a helicopter accident. Studio head Katzenberg left the company. The launch of the Euro Disney theme park had not gone well. In early 1995, Eisner met with investors and reassured them that Disney was well positioned to thrive moving forward. He recommitted to the 20 percent growth target. Equity analyst Jeffrey Logsdon reported Eisner's attitude as follows: "I'm not going to be the guy who breaks that (growth) streak."[24]

The second half of Eisner's tenure did not go smoothly. A series of animated movie releases failed to dazzle at the box office. Eisner needed to make bold bets to find the organic growth he desired. When new films and other products did not generate sufficient revenue, acquisitions became a more important part of his strategic arsenal. Troublesome deals included the $1.7 billion purchase of Infoseek. The company's stock price had outperformed the market by a wide margin during his first decade as CEO. Things changed starting in 1995, around the time that the company acquired Capital Cities/ABC. From the time that Disney announced that deal until Eisner stepped down, the stock price underperformed the S&P 500 Index by a substantial margin (see Figure 4.2). Revenue growth slowed to 5 percent per year in the final years of his tenure.

How ambitious was Eisner's goal of growing the company 20 percent per year from 1995 onward? That growth rate target implies that he intended to double the size of the company in less than four years. Imagine that! Walt Disney founded the company in 1923. It took 70 years for the company to reach $10 billion in sales. Eisner intended to generate another $10 billion of revenue in just four years. Audacious goals can be quite inspiring, but they also may induce costly strategic mistakes.

Many large firms find themselves in a similar situation, setting growth targets that become almost impossible to achieve once they have achieved a certain scale and scope. In 2001, Carol Loomis wrote an article for

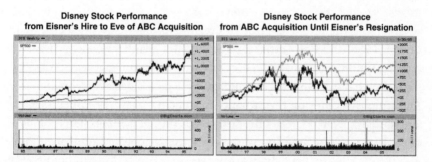

Figure 4.2 Walt Disney Company Stock Performance During the Michael Eisner Era

Source: Charts generated on www.bigcharts.com

Fortune magazine titled, "The 15% Delusion."[25] She noted that many large companies choose to establish an arbitrary target of growing earnings by 15 percent per year—doubling income every five years. Is that reasonable? She explained that after-tax corporate earnings grew only 8 percent per year over the last four decades of the 20th century. Loomis took a closer look at 150 of the largest firms in the United States. She discovered that only five of those companies grew earnings at a compound annual growth rate of 15 percent from 1980 to 1999. Loomis examined other time periods and found similar results. She questioned why chief executives continued to set such aggressive targets despite ample evidence that they were not achievable.

If you participate in the 15 percent delusion, then you place enormous pressures on those trying to create organic growth opportunities. A creative little niche business that will delight customers and deliver strong margins does not make the cut. Executives often won't fund those opportunities because they need much larger payoffs to achieve their goals. Enterprise leaders demand detailed forecasts of how large a new product or venture will become. How much capital investment is required? What share of the market can the venture secure? How much revenue and income will it generate? When will the venture achieve break-even status? Innovators typically do not receive funding unless they can provide a highly detailed return on investment analysis. They must demonstrate that a sizeable market opportunity exists. Time and again, creative ideas wither on the vine because of our obsession with precise projections.

Imagine the predicament that creative idea generators face in many organizations. If they estimate future sales prudently, they may not receive funding because their concept doesn't move the needle. On the other hand, if innovators engage in a bit of irrational exuberance, they might secure an investment, but run the risk of overpromising and underdelivering. At one consumer products firm where I conducted research, several innovative new items achieved tens of millions in revenue in the past few years, yet executives killed the projects because sales did not meet lofty expectations. One middle manager recounted to me:

We set up ourselves up to fail, by making promises that we could not keep. The same level of revenues might have been viewed as a success if we had set lower expectations. We couldn't do that though. Unless corporate believed it was a big opportunity, they would not provide us resources.

Niche versus Blockbuster

When executives demand to know whether an innovation will move the needle, they presume that innovators can distinguish blockbuster hits from niche products at the early stages of the product development process. Imagine how difficult that would be. When people search for and develop bold new ideas, they often cope with an incredible amount of ambiguity and uncertainty regarding customer needs, market conditions, and technological feasibility. The external environment proves turbulent, and accurate data can be hard to come by. Innovators test and probe amidst the fog, trying to ascertain whether their idea has merit and promise. Can we really predict what new products will move the needle?

Take the pharmaceutical industry, for example. Companies strive to develop blockbuster new drugs. The research and development process unfolds over many years. Researchers have examined whether analysts are able to generate accurate projections of product demand. Can they distinguish a blockbuster hit from a niche product during the research and development process? McKinsey researcher Myoung Cha and his colleagues studied this question using a dataset of 260 drug launches during the 2002–2011 time period. They examined more than 1,700 individual analyst forecasts. The McKinsey team discovered that, "More than 60% of the consensus forecasts in our data set were either over or under by more than 40% of the actual peak revenues."[26] Twenty percent of the drugs achieved peak sales at least 160 percent below analyst projections. While forecast accuracy improved as time passed, major errors remained even after the drug reached the market.

A few examples illustrate the dart-throwing nature of drug industry projections. Five years before the launch of Viagra, Pfizer projected peak annual sales of $500 million. One year prior to launch, outside analysts estimated that Viagra would reach $5 billion in revenue. What actually happened? Revenue peaked at $2 billion. For the cholesterol drug Lipitor, people estimated that sales would reach $500 million in five years. It actually achieved 10 times that revenue in its first five years on the market. Conversely, GSK's HIV drug Agenerase generated only $70 million in sales after three years, as compared to projections of $1.5 billion.[27]

Many companies struggle to predict product demand during the nascent stages of the creative process. One study of new ventures within large corporations found that first-year sales forecasts missed the mark by 80 percent.[28] Profit forecasts proved to be even more problematic. Still, most corporate resource allocation processes require people with a bold new idea to develop detailed revenue and profit projections at a *very* early stage. Entrepreneurial startups do not fare any better than big companies when it comes to forecasting revenues. In a recent study of German startups, Christopher Mokwa and Soenke Sievers discovered that sales projections missed the mark by 380 percent![29] Everyone appears to be throwing darts in the fog. Serial entrepreneur Steve Blank once said, "No one besides venture capitalists and the late Soviet Union requires five-year plans to forecast complete unknowns. These plans are generally fiction, and dreaming them up is almost always a waste of time."[30]

Time to Ripen

Sometimes, a creative idea takes time to flourish. The market opportunity for the new venture might seem small at first. People may conclude that the concept will appeal to a very narrow target market and occupy a profitable, but tiny, niche in the industry. No one foresees broader appeal to a mainstream audience for this radical concept. Many ideas take time to incubate. As creative individuals test and experiment, their ideas evolve and grow. It may take years for them to perfect the concept. Demanding precise forecasts, and only funding concepts that are expected to move the needle, may quash distinctive ideas with long incubation periods.

Imagine that an entrepreneur approached you with a concept for a new type of grocery store. He explains that the retailer will not provide many of the products and services typically offered by supermarkets. He proposes a store that will not run price promotions, accept coupons, or issue loyalty cards. The retailer will not advertise on television, issue circulars in the Sunday newspaper, or open any social media accounts. The store will stock very few national brands, focusing on private-label products instead. Product selection will be quite limited; the item count will be one-tenth of that found in most supermarkets. The product mix will not be stable; items will come and go rather frequently. The stores will be small, the aisles tight, and the parking lots quite crowded. The store will not offer self-checkout for customers. Finally, it will not offer a wide selection of fresh meat and fish.

Would you invest in this new grocery store concept? Admit it. You, like most people, would pass on that opportunity. The idea seems rather bizarre. What a mistake! Trader Joe's has become one of the most successful grocery retailers in the United States. In January 2018 the retailer ranked number one overall when over 11,000 American households were asked to rank their favorite supermarkets.[31] *Consumer Reports* repeatedly has ranked Trader Joe's as one of the best grocers in the nation as well. Sales per square foot consistently rank at the top of the supermarket industry (see Figure 4.3).[32]

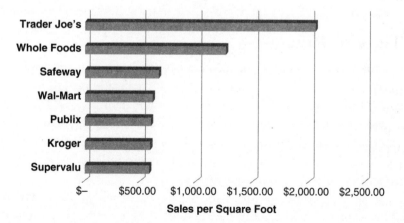

Figure 4.3   Grocery Sales per Square Foot for Selected Retailers in 2012

Source: Progressive Grocer's 2012 Super 50 Rankings (based on Nielsen data)[33]

The company has generated a cult-like following. Fans around the country cannot wait for a store to open in their area. Hundreds of people waited in line for hours at the grand opening of the retailer's Columbia, South Carolina, store. Local police had to direct traffic. Customers arrived at three o'clock in the morning for the grand opening of the Lexington, Kentucky, store. Fans across the country, such as Julie Merrill of Utah, have launched Facebook campaigns to convince the company to open a location in their area. Devoted followers such as Cherie Twohy have penned best-selling cookbooks featuring Trader Joe's products.[34] One person's unauthorized homemade commercial for Trader Joe's has attracted more than 1 million views on YouTube!

The success story did not emerge overnight though. Joe Coulombe opened his first Trader Joe's store in Pasadena, California, in 1967. The concept evolved slowly in those early years. He experimented with different products and merchandising strategies. One time he tried selling music albums, and on another occasion he stocked pantyhose in the stores. He discontinued both product lines. The company describes those years as the "still trying to find ourselves" era.[35] Coulombe began the shift to a private-label product strategy in 1972. He repositioned the firm to appeal more directly to "intelligent, educated, inquisitive individuals" who were health-conscious and enjoyed sampling new items.[36] In 1975 Coulombe began to cut and wrap cheese in the stores, and he introduced a ship's bell as a means of communicating key messages to employees (as opposed to an intercom system).

The retailer typically opened one new store per year in those days. It took 22 years for the chain to reach 30 store locations with total revenues of $150 million. The company did not expand beyond California until the early 1990s. Trader Joe's did not start selling its Charles Shaw wines (commonly known as "Two-Buck Chuck") until 2002. Today, the retailer has more than 470 store locations around the country, and it generates over $11 billion in revenue per year.[37]

Imagine if someone asked Joe Coulombe back in 1967 whether his concept would move the needle. Could he have predicted that the firm would blossom beyond the small niche it occupied in Southern California? Did he have any idea how long it would take to perfect the concept and accelerate growth? To his credit, Coulombe did not worry much about long-term projections. He listened to customers, catered to their needs and desires, and tinkered quite a bit with his strategy in those early years. While others might have spent time trying to forecast the future, he remained laser-focused on creating an amazing in-store experience. He built a core group of passionate fans this way. Word-of-mouth spread slowly at first, and eventually, interest swelled in the grocery store that looked and felt like no other.

The lesson is simple: Stop worrying about the forecast of how big your opportunity might be a few years from now. Instead, devote all your energy to delighting customers today, tomorrow, and the next day. Listen, observe, and learn from them at every opportunity. The results will come if you create an unparalleled end-to-end customer experience. Before you know it, a niche product or service might just become an unexpected nationwide phenomenon if you do your job exceptionally well.

Endnotes

1. Andrew Feinberg, "Booyah! The Manic Universe of Jim Cramer," *Kiplinger*, August 31, 2006 (www.kiplinger.com/ article/investing/T052-C000-S002-booyah-the-manic-universe-of-jim-cramer.html, accessed November 14, 2017).

2. Jonathan Hartley and Matthew Olson, "Jim Cramer's 'Mad Money' Charitable Trust Performance and Factor Attribution," May 23, 2016 (www-stat.wharton .upenn.edu/~maolson/docs/cramer.pdf, accessed November 14, 2017).

3. "Good Will Hunting," Box Office Mojo, n.d. (www.boxofficemojo.com/ movies/?id=goodwillhunting.htm, accessed December 11, 2017).

4. Samantha Miller, "Sexiest Man Alive: Ben Affleck," *People*, December 2, 2002 (people.com/archive/cover-story-sexiest-man-alive-ben-affleck-vol-58-no-23/, accessed December 11, 2017).

5. Chris Schulz, "The Terrible Film Festival: Gigli Is Even Worse Now Than It Already Was," *New Zealand Herald*, July 19, 2017, (www.nzherald.co.nz/ entertainment/news/article.cfm?c_id=1501119&objectid=11892008, accessed December 11, 2017).

6. Xan Brooks, "The 10 Biggest Box Office Flops of All Time – In Pictures," *The Guardian*, March 20, 2012 (www.theguardian.com/film/gallery/2012/mar/20/ biggest-box-office-flops-in-pictures, accessed December 11, 2017). Detailed cost and revenue data from Box Office Mojo.

7. "Raiders of the Lost Ark," Box Office Mojo, n.d. (www.boxofficemojo.com/ movies/?id=raidersofthelostark.htm, accessed December 11, 2017).

8. William Goldman, *Adventures in the Screen Trade* (New York: Warner Books, 1983), 41.

9. Ben Fritz, "How 'Wonder' Became a Box-Office Hit," *Wall Street Journal*, December 6, 2017 (www.wsj.com/articles/how-wonder-became-a-box-office-hit-1512577852, accessed December 11, 2017). Revenue numbers from Box Office Mojo.

10. Goldman, *Adventures in the Screen Trade*, 42.

11. Data compiled by author using information from Pro Football Reference (www.pro-football-reference.com/, accessed December 12, 2017).

12. Tony Manfred, "The Baltimore Ravens Have A Brilliant New Philosophy on the NFL Draft," Business Insider, May 5, 2014 (www.businessinsider.com/baltimore-ravens-nfl-draft-philosophy-2014-5, accessed December 12, 2017).

13. Pro Football Reference.

14. Phil Tetlock and Dan Gardner, *Superforecasting: The Art and Science of Prediction* (New York: Crown Publishing, 2015), 4.

15. Tetlock and Gardner, *Superforecasting,* 2015.

16. Tara Elizabeth Burton, "Could You Be a 'Super-Forecaster'?" BBC Future, January 20, 2015 (www.bbc.com/future /story/20150120-are-you-a-super-forecaster, accessed December 14, 2017).

17. Tetlock and Gardner, *Superforecasting,* 2015.

18. Arie Kruglanski and Donna Webster, "Motivated Closing of the Mind: 'Seizing' and 'Mreezing'," *Psychological Review*, 103(2), 1996, 263–283.

19. Jennifer Whitson and Adam Galinsky, "Lacking Control Increases Illusory Pattern Perception," *Science,* 322(5898), 2008, 115.

20. Nicholas Turiano, Benjamin Chapman, Stefan Agrigoroaei, Frank Infurna, and Margie Lachman, "Perceived Control Reduces Mortality Risk at Low, Not High, Education Levels," *Health Psychology*, 33(8), 2014, 883–890.

21. Ellen Langer, "The Illusion of Control," *Journal of Personality and Social Psychology*, 32(2), 1975, 311–328.

22. Whitson and Galinksy, 117.

23. This section draws upon a case study that I have taught for many years. Michael Rukstad and David Collis, "The Walt Disney Company: The Entertainment King," Harvard Business School Case Study 9-701-035, January 5, 2009. This section also draws upon what I learned when I spent time with Jim Fielding and his management team during his tenure as President of Disney Stores.

24. Tom Petruno, "A Bold Bit of Disney Image-ineering Pays Off," *Los Angeles Times,* January 30, 1995 (articles.latimes.com/1995-01-30/business/fi-26137_1_disney-world, accessed December 15, 2017).

25. Carol Loomis, "The 15% Delusion," *Fortune*, February 5, 2001 (archive.fortune.com/magazines/ fortune/fortune_archive/2001/02/05/296141/index.htm, accessed December 15, 2017).

26. Myoung Cha, Bassel Rifai, and Pasha Sarraf. "Pharmaceutical Forecasting: Throwing Darts?" *Nature Reviews. Drug Discovery*, 12(10), 2013, 737.

27. Harlan Sonderling, "Forecasting New Drug Sales Is More Art Than Science," *Business Insider*, April 14, 2014 (www.businessinsider.com/predicting-new-drug-sales-2014-4, accessed December 18, 2017).

28. Carmen Nobel, "Why Companies Fail and How Their Founders Can Bounce Back," Harvard Business School Working Ideas, March 7, 2011 (hbswk.hbs.edu/item/why-companies-failand-how-their-founders-can-bounce-back, accessed December 18, 2017).

29. Christopher Mokwa and Soenke Sievers, "Biases in Management Forecasts of Venture-Backed Start-Ups: Evidence from Internal Due Diligence Documents of VC Investors" July 11, 2012 (papers.ssrn.com/sol3/papers.cfm?abstract_id=1714399, accessed December 18, 2017).

30. Steven Blank, "Why the Lean Startup Changes Everything," *Harvard Business Review*, May 2013, 68.

31. Ashley Stewart, "Which Grocers Do Shoppers Prefer?" *Puget Sound Business Journal*, January 18, 2018 (www.bizjournals.com/bizwomen/news/latest-news/2018/01/which-grocer-do-shoppers-prefer.html, accessed March 11, 2018).

32. David Ager and Michael Roberto, "Trader Joe's," Harvard Business School Case Study 9-714-419, April 8, 2014.

33. Ibid.

34. For example, see Cherie Mercer Twohy *The I Love Trader Joe's Cookbook* (Berkeley, CA: Ulysses Press, 2009).

35. Trader Joe's, "About Trader Joe's: Timeline," Trader Joe's corporate website (www.traderjoes.com/pdf/tjs-timeline.pdf, accessed December 19, 2017).

36. Mark Gardiner, *Build a Brand Like Trader Joe's* (Seattle: Amazon Digital Services, 2012).

37. Trader Joe's corporate website (www.traderjoes.com/, accessed December 19, 2017). https://buffalonews.com/2018/05/10/wegmans-tops-make-list-of-top-selling-grocers/

CHAPTER 5

The Structural Mindset

We tend to meet any new situation by reorganization, and a wonderful method it is for creating the illusion of progress at a mere cost of confusion, inefficiency and demoralization.
　　　　　　—Charlton Ogburn, Jr., journalist and author

Superhero posters adorn the walls and desks in the "Hall of Justice," otherwise known as the corporate legal department. The internal walls of cubicles have been removed, leaving only the exterior frames in the communications team's office space. The odd layout signals the group's mission: We promote transparency. As you enter the life coach's office, you may choose to sit on a royal throne, wear a crown, and have your photo taken. Welcome to Zappos, the online shoes and apparel retailer known for its exceptional customer service, quirky culture, and imaginatively decorated headquarters.[1]

Zappos empowers its employees to do whatever it takes to please the customer. Individuals answering the phone in the customer service center do not follow scripts and do not get rewarded for closing out calls quickly. Instead, they engage in genuine, sometimes free-wheeling, conversation with their customers. One such call lasted more than 10 hours! Employees have the freedom to go the extra mile for customers, and the stories of heartwarming good deeds are legendary inside the firm. For instance, a customer service representative once fielded a call from a woman whose husband had died suddenly. The customer asked about returning shoes that she had just ordered that her spouse had never had the opportunity to wear. The Zappos employee refunded the money promptly and, without seeking permission from her boss, sent the woman flowers as an expression of sympathy. The customer recounted the story of generosity and compassion at her husband's funeral. You might think that this story is unique. Actually, the company sends hundreds of "wow" gifts to customers each year, including flowers and cards. This type of service has generated top-notch customer satisfaction and immense loyalty. Returning customers account for a large share of the firm's revenues.[2]

Tony Hsieh became the CEO of Zappos in 2000, shortly after Nick Swinmurn founded the online retailing startup. Within nine years, he had grown Zappos to more than $1 billion in revenue and sold the firm to Amazon. In 2012, Hsieh attended a conference in Austin, Texas, and heard from Brian Robertson, an avid proponent of self-management. Robertson developed the concept of Holacracy, a system that abolishes traditional hierarchies in the workplace.[3] Hsieh chose to embrace this egalitarian philosophy at Zappos. It seemed to fit with his attitudes regarding empowerment and collaboration. Hsieh eradicated the old organization structure, as well as the titles that came with it. People no longer serve as managers with stable teams of subordinates reporting to them. Instead, people function as *lead links* of self-organizing groups, known as *circles* in the nomenclature of Holacracy. These lead links do not have the formal authority and the titles associated with bosses in a traditional hierarchy. People do not have jobs; they occupy multiple *roles* at Zappos and belong to multiple circles. How exactly does this new system work? Robertson specifies how Holacracy should function in a written constitution, complete with a host of rather complicated rules and processes for how organizations should implement his self-management philosophy.[4]

Hsieh took the experiment one step further in the spring of 2015, embracing further moves toward self-organization. Inspired by the book, *Reinventing Organizations* by Frederic Laloux, Hsieh decided to eliminate the remnants of the hierarchical structure that still remained.[5] He penned a lengthy memo to employees, explaining that it was time to "rip the Band-Aid" and embrace self-management more completely:

> We've been operating partially under Holacracy and partially under the legacy management hierarchy in parallel for over a year now. Having one foot in one world while having the other foot in the other world has slowed down our transformation towards self-management and self-organization ... As of 4/30/15, in order to eliminate the legacy management hierarchy, there will be effectively no more people managers.[6]

The transition to self-management did not proceed smoothly, to say the least. Employees became confused and frustrated by the new organizational system. Some former managers struggled with the loss of control and the ambiguity concerning their new roles and responsibilities. Many employees chafed under the strict rules governing meetings and communication, which seemed ironic given that Holacracy evangelists claimed that the system was more empowering than traditional hierarchies.[7]

Hsieh offered employees a severance package in 2015 if they chose not to embrace this philosophy. In the year that followed, nearly one in five of the company's employees met the designated conditions and chose to accept the buyout. Another 11 percent departed without the buyout.[8] Employee dissatisfaction caused Zappos to tumble in *Fortune's* annual "Best Companies to Work For" list, which draws on worker surveys to evaluate organizations.[9] Once ranked as high as number 6 on the list, Zappos fell to number 86 in 2015.[10] In the following year, Zappos fell off the list completely for the first time since the previous decade. According to *Fortune*: "Two questions that generated particularly dismal results: Do employees think management has 'a clear view of where the organization is going and how to get there?' And do managers 'avoid playing favorites?'"[11]

While Zappos has persisted with the move to self-management, other firms have abandoned Holacracy after valiant attempts to make it work. Medium, the online publishing platform created by Evan Williams, chose to step away from its experiment with Holacracy in early 2016. The company had spent several years trying to implement this self-management system. Medium's director of operations, Andy Doyle, explained that, "Our experience was that it was difficult to coordinate efforts at scale . . . for larger initiatives, which require coordination across functions, it can be time-consuming and divisive to gain alignment."[12] *Fortune's* Jennifer Reingold studied both Zappos' and Medium's attempts to implement self-organization. She concluded:

> Holacracy is supposed to make work, well, work . . . And yet—as I saw at Zappos—the sheer number of rules and regulations, combined with the potential for politics to seep in in different forms, makes holacracy, in my view, a questionable replacement for the classic management system, as flawed as the latter may be . . . It may simply be, to borrow a phrase, that old-fashioned hierarchy is the worst approach—except for every other one that has been tried.[13]

Does Hierarchy Help or Hurt?

Like Brian Robertson and Tony Hsieh, many leaders and consultants have proclaimed hierarchy's evils over the years. They have argued that hierarchy stifles creativity, lowers employee engagement, and makes it difficult for people at lower levels to express dissenting views. These same voices typically advocate flatter organizational structures, even if they do not go so far as to embrace Holacracy. Less hierarchy translates into more creativity and innovation according to longstanding conventional wisdom. These arguments have a good deal of merit. They don't tell the whole story though.

A recent stream of research has demonstrated that hierarchy does have benefits that should not be overlooked. After reviewing this work, management scholars Bret Sanner and J. Stuart Bunderson argued that the appropriately designed hierarchy can promote organizational learning and innovation, contrary to conventional beliefs about optimal organizational structures. They even cite biological research on animal behavior and explain that many species organize themselves hierarchically, with important benefits for how groups of animals perform key tasks successfully. For instance, hierarchies reduce the costs and risks associated with fighting for resources, enhancing population stability in certain species. Sanner and Bunderson concluded that:

> Hierarchies help teams of people innovate much the same way they help animals survive in the wild—they keep teams moving in the same direction even when strong disagreements threaten to keep the teams from progressing or even tear them apart.[14]

How might we explore the potential benefits of hierarchy? What if we compared groups with a well-defined hierarchy to those filled with many high-status individuals? On the one hand, competition among multiple superstars might be counterproductive. On the other hand, well-defined social structures may lead to more efficient performance because they reduce conflict and enhance coordination among group members. Clear hierarchies may reduce role ambiguity and confusion within groups. Imagine yourself working on a project for a moment. If each person on the team thinks that he or she is the alpha dog, what will happen? You probably have witnessed a few battles between head-strong individuals in these situations, and in those cases, performance probably suffered.

Richard Ronay and his colleagues designed several ingenious experiments to compare the performance of egalitarian versus hierarchical groups. The researchers created three types of groups for their first experiment. In one set of groups, all members had high levels of power. Other teams consisted entirely of "low-power" members. Finally, they created groups with clearly defined hierarchies (high-, medium-, and low-power participants). They gave these groups two tasks to perform, one of which required them to work much more interdependently than the other.

What did the scholars discover? For tasks requiring little coordination, hierarchical groups did not perform any better or worse than groups consisting of all low-power or high-power members. Things changed when the groups needed to coordinate and integrate their actions to complete the task. The groups with a defined pecking order outperformed those consisting of all high-power or all low-power members.

In a second experiment, Ronay and his colleagues chose to examine the biological basis for hierarchical structures. They used a well-known proxy to measure individual differences in prenatal testosterone exposure. Again, they created three types of groups. One set of groups consisted of all high-testosterone members, while other groups only contained individuals with low levels of prenatal testosterone exposure. Finally, a third set of groups consisted of people with varying levels of testosterone exposure. For this experiment, they measured intragroup conflict while groups performed the assigned tasks. The hierarchical groups once again performed best. Moreover, the teams consisting of all high-testosterone members experienced much more conflict than the other groups, suggesting an important cause for the lower performance of these groups.[15]

How does this dynamic play out in teams outside of the laboratory? Nir Halevy and his co-authors examined National Basketball Association teams across 11 seasons. The researchers compared teams with wide dispersions amongst the players in terms of salary and time spent in the starting lineup to those with more egalitarian structures. The teams with a more pronounced pecking order won more games than those with smaller dispersions in pay and games spent as a starter. Why? Halevy and his co-authors hypothesized that the well-defined hierarchies tend to exhibit more coordination and cooperation among the players. They measured coordination by tracking statistics such as assists and turnovers. For instance, assists represent passes made from one teammate to another, helping a fellow player score a basket. Assists clearly represent a cooperative behavior rather than a selfish action. The data suggest a striking causal path. Wider dispersions of pay led to more cooperative behavior, ultimately leading to more wins.[16]

Think about the 1995–1996 Chicago Bulls, a team that won 72 of 82 games in the regular season. That team had a clear pecking order. Michael Jordan, considered by many to be the greatest player of all-time, led the team. Perennial all-star Scottie Pippen served as his very capable sidekick. Toni Kukoč provided scoring punch off the bench, earning Sixth Man of the Year honors. Role players such as Dennis Rodman (a rebounding specialist) and Steve Kerr (a three-point shooting specialist) rounded out the roster. The team finished in the top five in the league in terms of assists, and they recorded fewer turnovers than all but three teams that year.[17] They breezed through the playoffs and captured the NBA championship by defeating the Seattle Supersonics.

Should we conclude that hierarchy is always a good thing? Have we been listening to faulty arguments about the benefits of flatter organizations? Not necessarily. A clear pecking order does not always pay off in sports. Matt Bloom has shown that wider pay dispersions in major league baseball lead to lower team performance, a clear contrast to the findings in the NBA.[18] Why the difference between the two sports? Interdependence. In baseball, the players do not have to coordinate and integrate their actions nearly as much as basketball teammates do. Superstar starting pitcher Clayton Kershaw of the Los Angeles Dodgers can do his thing on the mound, and his actions have virtually nothing to do with how well slugger Cody Bellinger performs at the plate. Therefore, costs of a well-defined pecking order may outweigh the benefits in major league baseball. For instance, baseball players may perceive substantial salary differences as unjust, and those negative emotions could diminish performance.

Mountain climbers perhaps provide the most interesting setting for understanding the benefits and costs of hierarchy. Years ago, I studied the 1996 Mount Everest tragedy, in which accomplished mountaineers Rob Hall and Scott Fisher died while leading expeditions on the world's tallest mountain.[19] Three of their clients died as well. In that case, the status hierarchy within the teams appeared to diminish people's willingness to speak up and voice their concerns. One climber described a "clear guide-client protocol."[20] As a client, he explained that, "We had been specifically indoctrinated not to question our guides' judgment."[21] Even one of the guides felt pressured not to speak up because of the status hierarchy. Neil Beidleman reflected, "I was definitely considered the third guide . . . so I tried not to be too pushy. As a consequence, I didn't always speak up when maybe I should have, and now I kick myself for it."[22] The lack of open dialogue and dissent contributed to faulty decision making. Many climbers found themselves continuing toward the summit long after they should have turned around. They ignored their own rules, yet no one appeared to question the decision to do so.

Years after my study, Eric Anicich and his colleagues assembled a dataset on Himalayan expeditions stretching back for one hundred years. They identified over 5,000 expeditions consisting of more than 30,000 climbers. These expeditions included people from many different countries, allowing the researchers to compare teams from more hierarchical cultures with those from more egalitarian societies. What did they find? On the one hand, more climbers reached the summit on the teams from hierarchical cultures. On the other hand, more climbers died on these expeditions as well.[23] Why? Just as in the 1996 Everest expedition, people at lower levels of the pecking order may be afraid to voice concerns and, therefore, risky decisions may go unchallenged. Hierarchy, in this case, might help you achieve a lifelong goal, but it comes with potentially tragic costs.[24]

The Obsession with Reorganizations

Tony Hsieh may not have recognized the benefits of hierarchy, or acknowledged the disruptive effects that implementing Holacracy would unleash. Perhaps he failed to comprehend that taking any organizing system to the extreme often proves unproductive. Most importantly, though, he presumed that organizational structure drives performance, as many business leaders do. Unfortunately, that causal link is much more complex than many executives realize, as the studies of mountain climbers and sports teams illustrate. Leaders can adopt a variety of organizational structures, and each comes with its own costs and benefits. We cannot simply crank up an algorithm and select an optimal structure that promotes creativity, innovation, and growth. No such perfect structure exists, no matter the strategy, industry, or circumstance.

Business leaders do not seem to recognize this eternal truth. They desperately seek to find the perfect organization chart for their business. Executives love reorganizations. A substantial percentage of chief executives restructure their organizations early on in their tenure.[25] Like mad scientists in the laboratory, they invent new ways of dividing responsibilities. Leaders redraw the boxes and arrows on the organization chart, sometimes with a series of dreaded dotted lines to boot. Perhaps they decide to organize by line of business rather than by functional area, in hopes of breaking down discipline-based silos. One executive might choose to impose more centralization, perhaps in hopes of securing cost synergies. Another may decide to decentralize, in hopes of generating growth by providing more autonomy to local managers. Some add management layers; others eliminate them.

Sony provides a classic example of a reorganization aimed at unleashing employee creativity and recharging growth. The firm attempted a major restructuring in 2009. Apple had surpassed the Japanese consumer electronics giant in the portable music player business, and Samsung had overtaken the company in the flat-screen television industry. Once heralded as an innovative leader, Sony had become a stodgy laggard. At the time of the restructuring, the company's press release stated explicitly that the goal was to improve the pace and quality of new product development. CEO Howard Stringer proclaimed:

> This reorganization is designed to transform Sony into a more innovative, integrated and agile global company with its next generation of leadership firmly in place. The changes we're announcing today will accelerate the transformation of the Company that began four years ago.[26]

Stringer even appointed former IBM executive George Bailey to the newly created position of chief transformation officer.[27]

Three years later, Stringer stepped down as CEO. The company posted a net loss in each year from 2009 through 2012. Revenue dropped by 16 percent during that period. Sony fell further behind industry leaders in the consumer electronics business. The company's stock price rose in the immediate aftermath of Stringer's restructuring, but then dropped nearly 50 percent in his final two years as chief executive.[28] Kazuo Hirai took over as CEO in 2012. He began by – you guessed it – announcing a major reorganization.[29]

Some executives reorganize often, confusing and befuddling employees at all levels. Workers find themselves suddenly reporting to new managers and working with new colleagues. Who has authority in certain circumstances? Who has the decision rights on a particular issue? Often people struggle to answer these questions amidst frequent reorganizations. Productivity and creativity suffer as people confer frequently at the water cooler, trying to make sense of management's latest pronouncements. Fortunately, some veteran employees just put their heads down and plow ahead. They mutter to themselves *this too shall pass*, recognizing that the boxes and lines on the chart might change before the ink even dries.

Does all this reorganization add value? Probably not. One recent McKinsey and Company study concluded that only 16 percent of restructurings delivered unequivocally effective results.[30] Bain and Company analyzed over 50 major reorganizations during a six-year period, and the consultancy concluded that less than one-third generated improved financial results.[31] Many large companies have embraced flatter organizations in hopes of stimulating creativity and innovation, yet faster growth has not materialized. Still, boxes and lines keep shifting. Wharton's Peter Cappelli compares serial reorganizing to prescribing antibiotics very frequently for minor infections. You might alleviate the pain at that moment but harm the patient over time. He explains that trust in top management deteriorates as repeated reorganizations take place.[32]

Why do leaders keep trying to stimulate innovation by changing the organizational structure, despite the decidedly mixed results from prior efforts? The answer is straightforward: Leaders find it relatively easy to redraw the boxes and arrows on an organization chart. A shift in responsibilities here, a new dotted-line relationship there, the removal of a layer or two, and *voilà!* No matter how hard they search, leaders will not find an optimal structure that unleashes creativity. No such perfect solution exists. You cannot find a simple causal path that connects structure to performance.

Upon what dimensions of the organization should executives focus their energies if they wish to stimulate creativity? Leaders need to think about how teams perform their work, and how they can create the conditions that will enable those groups to flourish. The best leaders pay close attention to team climate, behavioral norms and ground rules, and the design of the work itself (see Figure 5.1). These three factors take time to develop appropriately, as well as a great deal of concerted effort. Developing the right climate, defining the ground rules, and designing the work proves much more difficult than redrawing boxes, lines, and arrows.

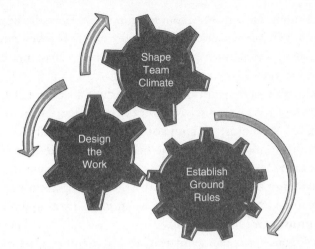

Figure 5.1   Building Work Environments to Stimulate Creativity

Shaping Team Climate

Several years ago, Google set out to determine what distinguished the rock-star teams at the company from lower performing groups.[33] Julia Rozovsky and her colleagues sought to identify teams where *the whole is greater than the sum of its parts*. Naming the initiative after the Greek philosopher attributed with this concept, they dubbed the initiative Project Aristotle. Rozovsky's People Analytics team collected data on 180 teams throughout Google—115 teams from engineering and 65 teams from sales.[34] They studied employee survey results over many years, and they conducted hundreds of interviews. Rozovsky and her colleagues also collected demographic information and data about the personality traits of team members. She explains what they discovered:

> We thought that building a perfect team would be pretty algorithmic in nature, because at Google, we love our algorithms. So we imagined that you just have to find the right number of superstars . . . you put them on a team, and *voilà*, you have your dream team assembled. We actually found something quite different at Google. What our research showed us was that it's less about who is on the team and more about how people interact that really makes a difference.[35]

The finding harkens back to a famous scene from the movie, *Miracle*, about the victorious 1980 U.S. Olympic Men's Hockey team. At one point, Coach Herb Brooks, played masterfully by Kurt Russell, sits high above the ice during the early days of tryouts. Craig Patrick, his assistant coach, approaches, and Brooks hands him a slip of paper with the names of the players he has selected for the team. Patrick expresses shock and dismay: "You're missing some of the best players." Brooks responds brusquely, "I'm not looking for the best players, Craig. I'm looking for the right ones."[36] The message was clear: A set of superstar individual athletes does not necessarily constitute a great team. Beating the vaunted Soviets at Lake Placid would require a whole that was much more than the sum of its parts.

Google, of course, had spent years trying to hire the best and brightest, often relying heavily on recruiting efforts at a few select institutions of higher education. Now their in-depth study suggested that talent did not distinguish the best teams from the rest at Google. What did matter most? The People Analytics group identified five attributes of the highest-performing teams. Of these five factors, a climate of psychological safety proved to be the most important by far.[37]

The results may have shocked Google, but they did not surprise Amy Edmondson, a scholar at Harvard Business School. More than two decades earlier, Edmondson developed the concept of team psychological safety. She defines it as the "shared belief held by members of a team that the team is safe for interpersonal risk taking."[38] Early in her academic career, Edmondson examined drug-related errors at two hospitals. She found that the teams with more reported errors tended to have nurse managers who exhibited better coaching skills and who led teams with higher-quality relationships among their members. How could better leadership lead to more mistakes? Perhaps that is the wrong question. Not surprisingly, the more effective leaders made it safe for people to discuss medical errors. Better teams do not make more errors; they simply are more willing to talk about and learn from them.[39] This led to their higher reporting rate. Over the course of many other rigorous studies, Edmondson has found that learning flourishes and team performance increases when people feel safe speaking up, sharing ideas, admitting mistakes, and asking questions.[40]

Our research together on the *Columbia* space shuttle accident demonstrates how low psychological safety can have tragic consequences by preventing people from ever having an opportunity to employ their creative problem-solving capabilities. After the foam strike occurred during launch in January 2003, some engineers feared the worst. They worried that the debris may have punctured a sizeable hole in the orbiter, which could cause a catastrophe when the shuttle tried to reenter the earth's atmosphere. Unfortunately, engineers did not feel safe expressing their concerns to senior managers. During the *Apollo 13* crisis decades earlier, a team developed an incredibly creative solution to a dangerous filtration problem using tube socks and duct tape. The solution saved the lives of Jim Lovell and his crew. For *Columbia*, such creative problem solving never took place, as management did not acknowledge that a serious safety risk existed. Therefore, senior leaders failed to empower a team to solve the problem.[41]

In the creative process, psychological safety enables team members to propose unconventional solutions, even ideas that might seem a bit crazy to others. People offer alternatives without worries about being ridiculed. Moreover, talking candidly about mistakes turns out to be absolutely critical during the creative process. Put simply, psychological safety enables more effective learning by doing, which is an essential characteristic of creative exploration. Prototyping and experimentation involve a healthy dose of failure. Early, low-fidelity prototypes will have many flaws. Some concepts will be rejected resoundingly by users. On many teams, people fear failure and worry about being blamed for poor results. Therefore, they will play it safe and not push the envelope.

Jason Park, Director of Innovation Strategy & Product Management at Allstate, recognizes that individuals will refrain from sharing early, low-fidelity prototypes if you do not create the right climate. He feels that adverse customer reactions do not worry most employees. Instead, individuals in some organizations worry about harsh criticism from their reporting managers and senior executives, and thus, they delay revealing their concepts time and again. Park believes that many people think to themselves, "I want to wait until I can show them the right answer." At Allstate, he encourages employees to take the risk and share a rough proposal with managers, even though the concept may be far from perfect. Adopting a common language around how "baked" a concept or proposal is (e.g., "preliminary," "discovery phase," etc.) can be useful to this end.[42]

How do leaders create a climate of high psychological safety? They must set a tone of openness to new ideas and hard truths (see Figure 5.2). Leaders must invite people to share proposals and conclusions that are not fully formed and assure them that negative repercussions will not ensue. Some leaders adhere to the old saying "Don't tell me about the flood. Build me an ark." In other words, don't come to me with problems unless you have all the answers. The best leaders make it safe to surface issues even if the solutions are not apparent. Christa Quarles, CEO of OpenTable, describes how she communicates the mantra of "early, often, ugly" to her team repeatedly. She says, "Early, often, ugly. It's O.K. It doesn't have to be perfect because then I can course-correct much, much faster. No amount of ugly truth scares me. It's just information to make a decision."[43]

Leaders must acknowledge that they are fallible to promote psychological safety. Showing some vulnerability can encourage others to offer seemingly radical ideas and come forward to talk about their mistakes in pursuit of innovation. Learning by doing flourishes in that type of

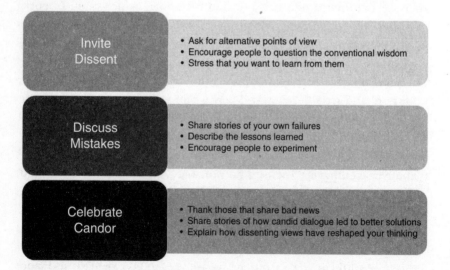

Figure 5.2   How Leaders Create Psychological Safety

atmosphere. Consider the case of Reed Hastings, Netflix founder and CEO. In 2011, he announced that the firm would divide its streaming business from the DVD unit, which he renamed Qwikster. The public revolted. Investors jeered. *Saturday Night Live* mocked him in a hilarious skit soon after the announcement.[44] To his credit, Hastings beat a hasty retreat and admitted his mistake very publicly. He talked openly about the fact that he had made errors many times during his career, and that he wanted everyone at Netflix to understand that mistakes represent powerful learning opportunities. Employees at Netflix heard the message loud and clear, and creativity continues to flourish at the company.[45]

When people speak up courageously, leaders must affirm them and celebrate their willingness to ask tough questions, talk about failures, and challenge the conventional wisdom. During his early days at Ford, former CEO Alan Mulally became frustrated when people were unwilling to acknowledge shortcomings in their units, despite the staggering losses at the automaker. Then one day, he applauded vigorously when an executive team member admitted that a project was "red," meaning a serious problem existed. He thanked him for being honest and asked, "How can we help you?" rather than admonishing the subordinate for poor performance or drilling him about the cause of the error. Soon others came forward with their own problems, and the senior team began to work collaboratively to solve them.[46]

Establishing Ground Rules

In baseball and softball, teams play in a wide variety of ballparks with different designs and dimensions. As a result, the umpire typically meets with coaches and captains before the start of every game to review the ground rules. The rules dictate how players and umpires should treat certain situations unique to that ballpark or field. For example, at Wrigley Field in Chicago, the umpire awards the batter second base if a batted ball becomes stuck in the ivy on the outfield walls. Fenway Park in Boston features an odd triangle-shaped section in deep right-center field. A yellow line stretches vertically up the wall above the triangle. If a ball strikes to the right of the line, the umpires award the batter a home run. If the ball strikes to the left, it remains in play. These ground rules ensure that players, coaches, and umpires are on the same page, so athletes can compete fairly and without confusion. As it turns out, the notion of ground rules can and should be applied to any team. Let's take a look at a few examples.

Pixar has produced eight Oscar-winning animated films including *Finding Nemo, Up, Toy Story 3,* and *Inside Out.* Fourteen of the studio's movies have grossed more than $200 million at the domestic box office. The films do not always start out beautifully, though. Ed Catmull, Pixar's longtime president, likes to say, "Pixar films are not good at first, and our job is to make them so—to go, as I say, 'from suck to not-suck.'"[47] How do they accomplish that feat? He credits a process that Pixar calls the "brain trust."[48]

The brain trust does not constitute a stable team. Instead, it represents the process through which a group of people with awesome storytelling chops come together periodically to review a film, ask questions, offer suggestions, and provide feedback. The brain trust began organically in the early days at Pixar. The process worked incredibly well for the original participants. When others tried to engage in this process, a few meetings became "flaming disasters."[49] Over time, Catmull realized that the brain trust could function more effectively if participants adhered to several important behavioral norms and ground rules. These principles could promote healthy dialogue and the candid exchange of ideas. When meetings broke down, as they inevitably did from time to time, Catmull could identify how participants failed to live by the rules of engagement and refocus their efforts in subsequent brain-trust sessions.

The brain trust operates according to four key rules of engagement.[50] First and most importantly, no one can override the director. They "remove the power structure from the room," meaning that the brain trust does not have the authority to mandate changes in the film. Even Catmull and other top executives do not force the filmmaker to enact changes. The director has the opportunity to assess all the feedback and make decisions about how to proceed. This ground rule ensures that the director will not enter the room in a defensive posture and stop listening to others' ideas. It requires restraint on the part of top leaders and a willingness for filmmakers to put themselves in a vulnerable position.

Catmull has learned that senior leaders, including himself, must hold back initially so as to encourage other participants to share their ideas.

Second, the process is peer to peer rather than a management review. Directors hear from fellow filmmakers who are working on other movie projects or shorts for the studio. People at all levels place themselves in the shoes of a storyteller rather than putting on their management hats. Third, they all have a vested interest in each other's success. Everyone commits to trying to be helpful to their colleague whose film is under review, rather than simply poking holes. The filmmakers participate in each other's brain-trust meetings, and that reciprocity becomes essential for creating trust and collegiality. Finally, each person commits to "giving and taking honest notes." In other words, candor is expected from all participants; no one should sugarcoat their feedback. Similarly, people should be open and willing to hear a wide variety of perspectives. The director strives to listen and learn, rather than explaining and defending each choice he or she has made. Filmmaker Andrew Stanton offers a wonderful metaphor for how the brain trust works. He describes Pixar as a hospital and its films as patients. The director is a doctor. The brain trust consists of fellow physicians trying to help the doctor cure a particular patient. Everyone wants the person to get well and they rally around that noble goal.

Toy Story 3 offers an example of the brain trust's positive impact on a film. During one meeting, Stanton offered a thought-provoking comment about the conclusion to the second act in the film. Filmmaker Michael Arndt had crafted a scene where Woody delivered an impassioned speech to provoke a mutiny amongst the day-care center toys in opposition to their dastardly boss, Lotso. Stanton remarked, "I don't buy it. These toys aren't stupid. They know Lotso isn't a good guy. They've only aligned themselves with him because he's the most powerful."[51] Eventually, Arndt came up with a new idea. He chose to focus on Lotso's enforcer, a doll named Big Baby. He asked himself, "What can Woody do that will turn Big Baby's sympathies against Lotso?"[52] Answering that question enabled Arndt to come up with a new "revolution" story that was more compelling and believable. The brain trust did not come up with the revised plot, but their questions stimulated Arndt to reframe the scene. He had been open to their comments, rather than reflexively defending the way he had crafted the mutiny.

Other firms employ behavioral norms and ground rules to guide creative processes as well. For instance, Doug Brown, President of UMass Memorial Community Hospitals, notes that the executive team for his healthcare system keeps a list of shared norms and ground rules up on the wall in the conference room that his team uses for weekly meetings. He finds that these norms are very useful to help the team engage in productive dialogue and debate. Brown notes that it is very important that team members hold the CEO and each other accountable for adhering to these rules of engagement. If someone doesn't behave in accordance with these norms, they can be gently, or perhaps not so gently, reminded about their unacceptable behavior. Brown believes that you have to be specific about the behaviors that you expect if you want teams to flourish. Moreover, you must foster peer-to-peer accountability, so that the leader does not always have to serve as the team's policeman.[53]

As you enter IDEO's offices, you will see rules posted on the walls as well. IDEO adheres to seven key rules during brainstorming sessions (see Table 5.1). These rules include: defer judgment, one conversation

Table 5.1   IDEO's Brainstorming Rules

Rule	Explanation
Defer Judgment	Withhold critique of others' ideas. Do not point out flaws or concerns.
Encourage Wild Ideas	Do not worry about practicality or feasibility at this point.
Build on the Ideas of Others	Employ the "yes, and" principle from improvisational comedy.
Stay Focused on the Topic	Concentrate on the "How might we?" question that initiated the discussion.
One Conversation At a Time	Avoid side conversations. Engage in active listening.
Be Visual	Describe your idea using simple sketches as well as words.
Go for Quantity	Strive to generate many ideas rather than only one or two high-quality concepts.

Source: Adapted from IDEO.com

at a time, and go for quantity rather than quality. These rules of engagement ensure that people do not squelch others' creative ideas, interrupt colleagues, or get distracted by side conversations. If people fail to live by these rules, facilitators step in to get the team back on track. Over time, of course, the rules have become second nature for the designers at the firm, much as they have for Pixar's filmmakers and Doug Brown's team at UMass Memorial.[54]

Designing the Work

Years ago, a group of young engineers at Data General accomplished the seemingly impossible task of designing a next-generation computer, dubbed the Eagle, in a very short period of time. Tracey Kidder chronicled their exploits masterfully in the prize-winning book, *The Soul of a New Machine*. This team thrived despite a rather unconventional boss and their own lack of industry experience. Why did they work so hard, and how did they come up with so many creative solutions? The engineers insisted that they did not do it for the money. Indeed, they did not receive hefty compensation for their work.

What motivated these engineers? One team member remarked that he strove for "opportunity, responsibility, and visibility" on this project.[55] These young people felt very strongly that they were working on a very important project. Most did not think that they could have been assigned to such a critical task at a large competitor, particularly given their youth and inexperience. As one team member explained, "At IBM we wouldn't have gotten on a project this good. They don't hand out projects like this to rookies."[56] The engineers understood the big picture too; they described their work as highly meaningful. Customers would benefit greatly from this innovation, and the company desperately needed a hit new product to survive. Individuals felt strongly that they did not want to let their colleagues down on such an important initiative. Team member Dave Epstein describes how he felt:

> I'm part of the team, and not only is my part very important, but so is Jim's part. Since Jim is killing himself—he's here every night until three in the morning—I would almost feel guilty if I wasn't working so hard, because I want it to be as much my project as it is his.[57]

Managers gave the engineers a great deal of autonomy. They set ambitious goals, but gave team members the freedom to determine the best way to accomplish those objectives. The engineers had the opportunity to apply a diverse array of personal skills and capabilities, too. The work seemed to change each week, as different challenges emerged. They certainly did not engage in repetitive, mundane work. As time progressed, the engineers felt very proprietary about the product. Steve Staudeher described that powerful sense of ownership: "I did that! I made that happen. That's the thrill, the final thrill."[58] Together, these features of the work created an environment in which intrinsic motivation soared.

The Eagle project team teaches us an important lesson about fostering creativity in our organizations. Perhaps, we should spend more time focusing on the design of the work, rather than the design of the organization chart. Richard Hackman and Greg Oldham developed a work–design model that helps us understand how we can foster intrinsic motivation and stimulate creativity on our teams (see Figure 5.3). These scholars identified five key job/task attributes, each of which was a critical factor on the Eagle project.

First, performance improves if the work requires people to utilize a variety of talents and skills. People exhibit stronger intrinsic motivation

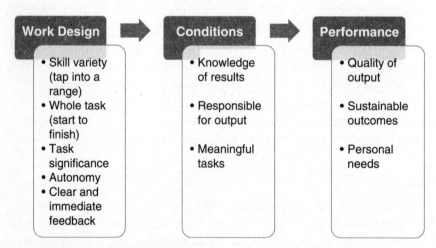

Figure 5.3   Designing the Work

Source: Adapted from Hackman and Oldham, 1976.[60]

if we challenge them in different ways over time. Second, workers thrive if they have the opportunity to see a task through from beginning to end. Results suffer if employees routinely work on small pieces of a larger puzzle, without an understanding of what the final result looks like. Third, outcomes improve if team members believe that their work will have a significant impact on the welfare and satisfaction of others. Fourth, workers thrive if provided substantial autonomy, such that they can make decisions independently about how best to perform the work. Finally, outcomes improve if workers receive timely and constructive feedback, so they know where they stand and can alter their behaviors as needed.[59]

Through their research, Hackman and Oldham demonstrated how designing the work properly enhances the extent to which people find their work meaningful. Moreover, it develops a sense of ownership and responsibility for the outcomes. If leaders design these five facets of the work appropriately, people experience greater levels of intrinsic motivation and produce higher-quality results. Individuals also become more committed to the team. Oldham's subsequent work specifically demonstrated that designing the work appropriately could have beneficial effects on creativity, measured in a variety of ways.[61]

The engineers on the *Eagle* project clearly found the work meaningful and felt a powerful sense of ownership. The team members exhibited high levels of intrinsic motivation, and the outcomes speak for themselves. The engineers on the *Eagle* project performed remarkably well on this creative task, and they beamed with pride when asked about their machine.

Stop Searching for a Panacea

Leaders need to create the setting in which people can excel. As former Harley Davidson CEO Richard Teerlink once said, "As a leader . . . your principal job is to create an operating environment where others can do great things."[62] Organization charts do not constitute the most important elements of an enabling environment. A safe climate, clear ground rules, and well-designed work; these elements do not come easily, but they provide the essential building blocks for creativity to flourish and for teams to thrive. Leaders everywhere need to stop searching for easy structural solutions. They will not find them, and may do more harm than good in pursuit of a panacea for their organizations.

Endnotes

1. This section draws from research that I conducted during a visit to Zappos headquarters several years ago. In case you are wondering, I did get my photo taken sitting on the royal throne with a crown on my head!

2. Micah Solomon, "Tony Hsieh Reveals the Secret to Zappos' Customer Service Success in One Word," *Forbes*, June 12, 2017 (www.forbes.com/sites/ micahsolomon/2017/06/12/tony-hsieh-spills-the-beans-the-one-word-secret-of-zappos-customer-service-success/#7c506e6d1acc, accessed January 3, 2018).

3. Holacracy, "How It Works," Holacracy.org (accessed December 27, 2017),

4. Jennifer Reingold, "How a Radical Shift Left Zappos Reeling," *Fortune*, March 4, 2016 (fortune.com/zappos-tony-hsieh-holacracy/, accessed January 3, 2018).

5. Frederic Laloux, *Reinventing Organizations* (Brussels: Nelson Park, 2014).

6. Rebecca Greenfield, "Zappos CEO Tony Hsieh: Adopt Holacracy or Leave," *Fast Company*, March 30, 2015 (www.fastcompany.com/3044417/zappos-ceo-tony-hsieh-adopt-holacracy-or-leave, accessed December 28, 2017).

7. For more on Zappos' Holacracy experiment, see Harvard Business School Professor Ethan Bernstein's interview on the subject (hbr.org/ideacast/2016/07/ the-zappos-holacracy-experiment.html); Tony Hsieh's comments about possible regrets(www.cnbc.com/2016/09/13/zappos-ceo-tony-hsieh-the-thing-i-regret-about-getting-rid-of-managers.html); Jena McGregor's article in the *Washington Post*: (www.washingtonpost.com/news/on.../wp/.../zappos-gets-rid-of-all-managers/); and Adam Grant's interview with Tony Hsieh for Knowledge@Wharton (knowledge.wharton.upenn.edu/article/zappos-tony-hsieh-holacracy-right-fit/)

8. Bourree Lam, "Why Are So Many Zappos Employees Leaving?" *The Atlantic*, January 15, 2016 (www.theatlantic.com/business/archive/2016/01/zappos-holacracy-hierarchy/424173/, accessed January 4, 2018).

9. Reingold, "How a Radical Shift Left Zappos Reeling".

10. "100 Best Companies to Work For," *Fortune*, February 2015 (fortune.com/best-companies/2015/, accessed January 6, 2018).

11. Reingold, "How a Radical Shift Left Zappos Reeling".

12. Andy Doyle, "Management and Organization at Medium," Medium.com, March 4, 2016 (blog.medium.com/management-and-organization-at-medium-2228cc9d93e9, accessed January 7, 2018).

13. Jennifer Reingold, "Management Changes at Medium," *Fortune*. March 4, 2016 (fortune.com/2016/03/04/management-changes-at-medium/, accessed January 7, 2018).

14. Bret Sanner and J. Stuart Bunderson, "The Truth about Hierarchy," *MIT Sloan Management Review* 50, Winter 2018.

15. Richard Ronay, Katharine Greenaway, Eric Anicich, and Adam Galinsky, "The Path to Glory Is Paved with Hierarchy: When Hierarchical Differentiation Increases Group Effectiveness," *Psychological Science,* 23(6), 2012, 669–677.

16. Nir Halevy, N., Eileen Chou, Adam Galinsky, and Keith Murnighan, "When Hierarchy Wins: Evidence from the National Basketball Association" *Social Psychological and Personality Science*, 3(4), 2012, 398–406.

17. I researched the team's performance using data available on www.basketball-reference.com/, accessed January 22, 2018.

18. Matt Bloom, "The Performance Effects of Pay Dispersion on Individuals and Organizations," *Academy of Management Journal,* 42(1), 1999, 25–40.

19. I have spoken in depth on multiple occasions with mountaineers Ed Viesturs and David Breashears about the May 1996 tragedy on Everest. Both Men were on the mountain at the time, as leaders of the IMAX expedition, and they helped with the rescue efforts when the Mountain Madness and Adventure Consultants teams encountered serious trouble. For more on my research regarding these events, see Michael Roberto, "Lessons from Everest: The Interaction of Cognitive Bias, Psychological Safety, and System Complexity," *California Management Review* 45(1), 2002, 136–158; Michael Roberto and Gina Carioggia, "Mount Everest – 1996," Harvard Business School Case Study 9-303-061, January 6, 2003.

20. Jon Krakauer, *Into Thin Air* (New York: Anchor Books, 1998), 245.

21. Krakauer, *Into Thin Air*, 245.

22. Krakauer, *Into Thin Air*, 260.

23. Eric Anicich, Roderick Swaab, and Adam Galinsky. "Hierarchical Cultural Values Predict Success and Mortality in High-Stakes Teams," *Proceedings of the National Academy of Sciences*, 112(5), 2015, 1338–1343.

24. The five deaths on the Mountain Madness and Adventure Consultants teams in May 1996 occurred during the descent from the summit (Rob Hall, Scott Fisher, Yasuko Namba, Doug Hansen, and Andy Harris).

25. Marcia W. Blenko, Michael Mankins, and Paul Rogers, "The Decision-Driven Organization," *Harvard Business Review,* June 2010, 54–62.

26. Sony, "Sony Corporation Announces Major Reorganization and New Management Team Led by Howard Stringer," press release, February 27, 2009 (www.sony.net/SonyInfo/News/Press/200902/09-028E/index.html, accessed January 22, 2018).

27. Sony, "Sony Corporation Names George Bailey Senior Vice President, Corporate Executive and Chief Transformation Officer," press release, May 15, 2009 (www.sony.net/SonyInfo/News/Press/200905/09-055E/index.html, accessed January 22, 2018).

28. Hiroko Tabuchi, "Incoming Chief Takes on a Sony That Is a Shadow of Its Former Self," *New York Times*, February 2, 2012(www.nytimes.com/2012/02/03/technology/incoming-chief-takes-on-a-sony-that-is-a-shadow-of-its-former-self.html, accessed January 22, 2018).

29. Hiroko Tabuchi and Bettina Wassener, "Sony Chief Unveils Plans to Revive Company," *New York Times*, April 12, 2012.

30. Stephen Heidari-Robinson and Suzanne Heywood, "Assessment: How Successful Was Your Company's Reorg?" *Harvard Business Review* (Digital Article), February 24, 2017.

31. Marcia W. Blenko, Michael C. Mankins and Paul Rogers, "The Key to Successful Corporate Reorganization," Forbes.com, July 30, 2010 (www.bain.com/publications/articles/key-to-successful-corporate-reorganization.aspx, accessed January 9, 2018).

32. "Another Reorganization? What to Expect, What to Avoid," Knowledge@Wharton, July 2, 2003 (knowledge.wharton.upenn.edu/article/another-reorganization-what-to-expect-what-to-avoid/, accessed January 10, 2018).

33. This section benefits from the lessons I have learned during several visits to Google over the past few years.

34. Google, "Guide: Understand Team Effectiveness," Re:Work (rework.withgoogle.com/print/guides/5721312655835136/, accessed January 11, 2018).

35. Google Partners, "Google's Five Keys to a Successful Team," June 22, 2015 (www.youtube.com/watch?v=KZlSq_Hf08M&t=1096s, accessed January 11, 2018).

36. *Miracle*, directed by Gavin O'Connor, Buena Vista Pictures, February 6, 2004.

37. Google Partners, "Google's Five Keys to a Successful Team."

38. Amy Edmondson, "Psychological Safety and Learning Behavior in Work Teams," *Administrative Science Quarterly*, 44(2), 1999, 350–383.

39. Amy Edmondson, "Learning from Mistakes Is Easier Said Than Done: Group and Organizational Influences on the Detection and Correction of Human Error," *Journal of Applied Behavioral Science*, 32, 5–32. Amy Edmondson, Anita Tucker, and I have co-authored a case study about a Minnesota hospital's attempt to speak more openly about medical accidents as a means of promoting learning and improvement. See Amy Edmondson, Michael Roberto, and Anita Tucker, "Children's Hospital and Clinics (A)," Harvard Business School Case Study 9-302-050, November 15, 2001.

40. Amy has written a terrific book on teams in which her work on psychological safety is featured prominently. See Amy Edmondson. *Teaming: How Organizations Learn, Innovate, and Compete in the Knowledge Economy* (New York: John Wiley & Sons, 2012).

41. Amy Edmondson, Richard Bohmer, and I have published a book chapter, journal article, and multimedia case study based on our research regarding the *Columbia* space shuttle accident. See Amy Edmondson, Michael Roberto, Richard Bohmer, Erika M. Ferlins, and Laura R. Feldman. "The Recovery Window: Organizational Learning Following Ambiguous Threats." In *Organization at the Limit: Lessons from the Columbia Disaster*, ed. William Starbuck and Moshe Farjoun, (New York: Wiley Blackwell, 2005), 220–245; Michael Roberto, Richard Bohmer, and Amy Edmondson, "Facing ambiguous threats," *Harvard Business Review*, November 2006, 106–113; Michael Roberto, Amy Edmondson, Richard Bohmer, Laura Feldman, and Erika Ferlins, "Columbia's Final Mission," Harvard Business School Multi-Media Case Study 9-305-032, March 1, 2005. Note that an updated version of this multimedia case study will be released in 2018.

42. Personal interview with Jason Park, July 2017.

43. Adam Bryant, "Christa Quarles of OpenTable: The Advantage of 'Early, Often, Ugly'," *New York Times*, August 12, 2016 (www.nytimes.com /2016/08/14/ business/christa-quarles-of-opentable-the-advantage-of-early-often-ugly.html, accessed January 15, 2018).

44. *Saturday Night Live*, "Netflix Apology," October 1, 2011.

45. Daniel B. Kline, "Netflix CEO Reed Hastings on the Beauty of Making Mistakes," *The Motley Fool*, September 26, 2015 (www.fool.com/investing/ general/2015/09/26/netflix-ceo-reed-hastings-on-making-mistakes.aspx, accessed January 15, 2018).

46. "Alan Mulally of Ford: Leaders Must Serve, with Courage," Stanford Graduate School of Business, February 7, 2011 (www.youtube.com/watch? v=ZIwz1KlKXP4, accessed January 15, 2018). Mulally has recounted this story many times. For a detailed written account of the episode, see Bill Vlasic, *Once Upon a Car: The Fall and Resurrection of America's Big Three Automakers—GM, Ford, and Chrysler* (New York: William Morrow, 2011).

47. Ed Catmull, *Creativity Inc.: Overcoming the Unseen Forces That Stand in the Way of True Inspiration* (New York: Random House, 2014), 90.

48. This section on Pixar draws upon a personal interview that I conducted with Ed Catmull in October 2014.

49. Ed Catmull, Talk from Fortune Brainstorm Tech Conference, July 14, 2015, Aspen, CO (www.youtube.com/watch?v=VMMKWVIUqm8, accessed January 16, 2018).

50. Ibid.

51. Catmull, *Creativity Inc.*, 101.

52. Catmull, *Creativity Inc.*, 101.

53. Personal interview with Doug Brown, January 2017.

54. Personal visit to IDEO's offices in San Francisco, March 2016.

55. Tracey Kidder, *The Soul of a New Machine* (Boston: Back Bay Books, 2000).

56. Ibid.

57. Data General interviews, Harvard Business School Video, 1996.

58. Ibid.

59. J. Richard Hackman and Greg Oldham. "Motivation through the design of work: Test of a theory," *Organizational Behavior and Human Performance*, 16(2), 1976, 250–279.

60. Ibid.

61. Greg Oldham and Anne Cummings, "Employee creativity: Personal and contextual factors at work," *Academy of Management Journal*, 39(3), 1996, 607–634.

62. Matt Mayberry, "20 Inspiring and Valuable Quotes on Leadership," Entrepreneur .com, October 27, 2015 (www.entrepreneur.com/article/252099, accessed January 17, 2018).

CHAPTER 6

The Focus Mindset

Indeed, I find that distance lends perspective and I often write better of a place when I am some distance from it. One can be so overwhelmed by the forest as to miss seeing the trees."

—Louis L'Amour, Novelist

Irish singer-songwriter Paul David Hewson, known to fans as Bono, described his feelings in this manner as U2 prepared to release its fourth studio album in 1984:

> I can't tell you where we're about to go but I know that I can't sleep at night with the thoughts of it. I'm so excited about this idea that we've just begun—the way I feel is that we're undertaking a real departure. I can't stop talking about it. It would take ten men to hold me down at the moment.[1]

The band had just completed the tour for its *War* album, which included the iconic protest song, *Sunday Bloody Sunday*. The Irish rockers achieved a commercial breakthrough with this album, as it reached number 1 in the UK and number 12 in the United States.[2] They began to play concerts at larger venues, and they gained a reputation for inspiring live performances.

Coming off this success, Bono and his bandmates could have rested on their laurels, but they chose to experiment instead. *Rolling Stone's* Steve Pond wrote that, "U2 saw itself in danger of becoming just another sloganeering arena-rock band."[3] The group hired Brian Eno and Daniel Lanois to produce a new album and to help take the group in a radical new direction. Bass guitarist Adam Clayton explained, "We were looking for something that was a bit more serious, more arty."[4]

The band members moved into Slane Castle in County Meath, Ireland, to write and record the songs for this album. They wanted to get away together and concentrate on their work.[5] Lanois had a history of taking bands to special places to record music, in hopes of stimulating the creative juices. He commented that:

> Bono was looking for a different kind of location, a building that had ghosts in the walls and some kind of a sense of history. So that we weren't just in an empty modern warehouse, that we were actually feeling the presence of goings-on from the past.

He explained that the band and crew developed camaraderie as they lived, worked, and relaxed in the 18th century castle. "The best part of it is that everyone was living there," Lanois said.[6] As a joke, they even performed songs while entirely naked one day. The group remained secluded in the castle for a month, recording songs in a library surrounded by books. In these writing and recording sessions, the band members tested some rather unconventional material.[7]

U2 titled the new album, *The Unforgettable Fire*, taking the name of a famous art exhibit about the atomic bombings of Hiroshima and Nagasaki at the end of World War II. Many critics felt that the album represented a creative rebirth for the band. Bono described it as if they had broken up and formed a new band with the very same musicians. The album reached number one in the UK, and the song "Pride (In the Name of Love)" became the band's first top 40 hit in the United States.[8]

The story of *The Unforgettable Fire* seems familiar to many of us. It conjures up visions of our favorite artists holed up somewhere, away from the busy routines of daily life, achieving creative breakthroughs. The Rolling Stones held a series of drug-fueled all-night recording sessions in a mansion on the French Riviera, culminating in the *Exile on Main St.* album, one of the band's greatest works.[9] Bob Dylan and The Band recorded *The Basement Tapes* album in the cellar of a home in Woodstock, New York.[10] More recently, Kanye West purportedly isolated himself on a Wyoming mountaintop for weeks to work on his newest album.[11]

Over the years, creative people in many fields have extolled the virtues of working in seclusion. Picasso once said, "Without great solitude no serious work can be done."[12] Inventor Nikolas Tesla explained, "Originality thrives in seclusion free of outside influences beating upon us to cripple the creative mind."[13]

Many of us continue to believe that we can accelerate creative work by giving a team the opportunity to concentrate, free of distractions, in a space that's all their own. In today's business world, that means sending people away to an off-site retreat, creating an innovation hub, or secluding a team in a war room for days or weeks on end. We all want our own Slane Castle.

The War Room at Google Ventures

In 2009 Google's Jake Knapp became frustrated by how the pace and structure of a typical day at work seemed to inhibit creative breakthroughs.[14] He recognized that he needed a better process for gathering a team of engineers to solve a tough business problem. He became obsessed with finding a more effective way to work. Knapp noticed the usual workday involved a series of interruptions and "context switches" that made it difficult to think clearly and collaborate effectively. He explained,

> Every meeting, email, and phone call fragments attention and prevents real work from getting done. Taken together, these interruptions are a wasp's nest dropped into the picnic of productivity . . . no doubt about it: fragmentation hurts productivity. Of course, nobody wants to work this way. We all want to get important work done. And we know that meaningful work, especially the kind of creative effort needed to solve big problems, requires long, uninterrupted blocks of time.[15]

Knapp developed the concept of a "design sprint" to solve the fragmentation problem that he had identified. He asked a group of people to clear their calendars for an entire week and to leave their electronic devices behind. He wanted them to focus all their energies on one problem for five days. He developed a detailed step-by-step process for how groups should use design thinking to solve a challenging problem. Day one involved defining the problem appropriately and consulting with key experts. Over the course of the week, the group developed alternative solutions, constructed storyboards, and built prototypes. At the end of the week, the team tested their ideas by soliciting feedback on their prototypes from users.

Knapp came to believe that a dedicated space enhanced the quality of a group's work. He advocated setting up a "war room" for a design sprint team. The room would become the group's home for five days. Knapp explained, "The walls of a war room can extend a team's memory, provide a canvas for shared note-taking, and act as long-term storage for works in progress."[16] He recommended moveable furniture and tons of whiteboards on the walls. The war room at Google Ventures became Knapp's Slane Castle. Over time, many other companies such as Facebook, Airbnb, Microsoft, and Uber have adopted the five-day design sprint methodology pioneered by Knapp.

The Dangers of Multitasking

Getting away and concentrating exclusively on a project has clear cognitive benefits. Still, many of us spend a great deal of our day attempting to multitask. Admit it. How many times have you been checking email, browsing the web, or reviewing a report while on a conference call? Or perhaps we have eaten our morning bagel and conducted a teleconference while driving to work. We have come to believe that we are superheroes, able to juggle many duties simultaneously with ease. In reality, we are fooling ourselves.

How often do we get interrupted at work? Gloria Mark and her colleagues decided to conduct in-depth observations of 24 workers at an information technology company.[17] They studied software developers, financial analysts, and managers. Mark explains that people switched activities every three minutes or so. Many situations involved self-interruptions rather than another worker approaching someone with a question or request. Some interruptions did not prove harmful, as they pertained to the same task and took little time. In other cases, though, the worker switched to an entirely different task. Mark found that it took more than 23 minutes for people to get back to their original work once they experienced an interruption.[18] She has found that these types of constant interruptions can be very stressful, thereby harming employee performance.

A great deal of research demonstrates that multitasking and interruptions can be highly detrimental to individual performance. Cyrus Foroughi and his colleagues have studied how interruptions affect the quality of work that we perform. They asked students to outline and write an essay. The scholars interrupted some students without warning several times, asking them to complete a different task for just 60 seconds on each occasion. Others worked on their essays without being disturbed. Foroughi and his co-authors asked two independent graders to read and grade the essays using the College Board's Essay Scoring Guide. The interrupted students received significantly lower evaluations on their essays.[19]

What about heavy media multitaskers, the kind of people who like to juggle multiple text message streams, email conversations, and the like. Eyal Ophir and his co-authors wondered whether some superstars had remarkable abilities to work on different things at the same time. In fact, they found that heavy media multitaskers consistently underperformed their more focused counterparts across several studies. The heavy multitaskers could not store and organize information more effectively than others, and they often failed to screen out extraneous data.[20]

Despite these research findings, many of us continue to believe that we possess superpowers. Unfortunately, our self-confidence often does not match our actual abilities. David Sanbonmatsu and his colleagues asked people about their driving habits. Did they use their cell phone while behind the wheel? They also inquired about people's beliefs regarding their multitasking capabilities. All participants in the study took a test to measure their actual ability to multitask effectively. The scholars found that the people who exhibit the most belief in their multitasking abilities, and who drive while talking on the phone most often, tend to perform more poorly on the test than others with less confidence.[21]

These findings seem to support Knapp's contention that we have to find dedicated time and space to concentrate on challenging problems.

We must avoid interruptions and stop trying to juggle so many different duties simultaneously. Many firms have acknowledged the dangers of work fragmentation and have tried to put creative teams in a better position to thrive. For that reason, these organizations send groups away for off-site retreats, set up war rooms or innovation hubs, and isolate project teams to concentrate on an issue. As it turns out, though, seclusion and concentration do not always pay off. Sometimes, we need a little distance from a problem to achieve a creative breakthrough. Unrelenting focus has its limits. We can get too close to a problem, and we can get stuck when we concentrate so intently on a challenging task.

When the Tank Runs Dry

British writer J. R. R. Tolkien published *The Hobbit* in 1937. Because of its astounding success, publisher Stanley Unwin asked Tolkien to write a sequel. He began work on *The Lord of the Rings* during the fall of 1937. Tolkien did not publish the trilogy's first book, *The Fellowship of the Ring*, until 1954. During the intervening years, Tolkien wrote intensely for some periods of time, but he took lengthy breaks from the project as well. He took time away from writing to sketch his characters and construct maps, even to write songs that the characters might sing.[22] He gathered informally on occasion with other writers, such as C. S. Lewis and Owen Barfield, at Oxford's Magdalen College, where Tolkien served on the faculty. These authors called themselves the Inklings, and they enjoyed discussing literature over a few pints of beer and sharing drafts of their work with one another. Tolkien often revised his manuscripts after giving his peers some time to consider what he had written and provide him feedback.[23]

When he was not working on the *The Lord of the Rings*, Tolkien embarked upon a series of other important projects. He performed his teaching duties as a professor, took on administrative obligations at Oxford, published scholarly essays, raised a family, and worked as an air raid warden during World War II. He drafted and illustrated letters to his children from Father Christmas, which were published posthumously, and he wrote a series of other short stories.[24] Tolkien even trained for a few days in 1939 at the top-secret Government Code and Cypher School, learning to crack Nazi codes, though he never worked for the British government in this capacity during the war.[25] Clearly, Tolkien's masterpiece came about through fits and starts. At times, his mind wandered off to other projects, only to return repeatedly to focus passionately on the trilogy. To date, readers have purchased more than 150 million copies of *The Lord of the Rings*, ranking them among the best-selling books in history.[26]

While Tolkien worked intermittently on his famous trilogy, Mark Twain's creative process sometimes involved an abrupt and rather lengthy break. Twain published *The Adventures of Tom Sawyer* in 1876. During the summer of that year, he traveled to upstate New York with his wife and children. They stayed for roughly three months at Quarry Farm, the home of his wife's sister and her husband. While vacationing there, Twain began work on a sequel titled *Adventures of Huckleberry Finn*.[27] He wrote a significant portion of the book that summer, but then he halted his writing in frustration. He considered tossing the draft completely. In a letter to a friend, Twain relayed his feelings about the work in progress: "I have written 400 pages on it—therefore it is very nearly half done. It is Huck Finn's Autobiography. I like it only tolerably well, as far as I have got, and may possibly pigeonhole or burn the MS when it is done."[28] Twain scholar Henry Nash Smith speculated that Huck Finn had become a much more complicated character than the author envisioned originally. He wrote, "It is as if the writer himself were discovering unsuspected meanings in what he had thought of as a story of picaresque adventure."[29]

In the next few years, Twain worked on several other projects. He published *The Prince and the Pauper* in 1881. In the following year, Twain embarked on a steamboat tour of the Mississippi. He departed from St. Louis on April 20th and eventually reached St. Paul, Minnesota, one month later.[30] He chronicled this trip, as well as his time as a steamboat pilot before the Civil War, in the book, *Life on the Mississippi*, published in 1883. Twain's steamboat trip produced more than this memoir, though. It inspired him to revisit the manuscript that he had abandoned at Quarry Farm and worked on quite sparingly during the intervening years.

Twain set about writing with single-minded intensity in the summer of 1883. He wrote to his friend, William Dean Howells, about his renewed interest in the story of Huck Finn as well as the furious pace of his writing:

> I have written eight or nine hundred manuscript pages in such a brief space of time that I mustn't name the number of days; I shouldn't believe it myself, and of course couldn't expect you to. I used to restrict myself to four or five hours a day and five days in the week, but this time I have wrought from breakfast till 5:15pm six days in the week, and once or twice I smouched a Sunday when the boss wasn't looking. Nothing is half so good as literature hooked on Sunday, on the sly.[31]

Twain finally published *Adventures of Huckleberry Finn* in December 1884, and many consider it one of the greatest American novels ever written. Interestingly, the seven-year interlude between the two furious bursts of writing did not represent the first time that Twain had stepped away from a manuscript only to revisit it with much success in later years. Twain recalled in his autobiography that he had stopped working on *The Adventures of Tom Sawyer* in 1873 and returned to it two years later. He learned about the value of taking a step back from a manuscript at that time. Twain explained:

> Ever since then, when I was writing a book, I have pigeonholed it without misgivings when its tank ran dry, well knowing that it would fill up again without any of my help within the next two or three years, and that then the work of completing it would be simple and easy.[32]

More than a century after Twain's instincts led him to "pigeonhole" his manuscripts, researchers have shown that taking a break indeed can stimulate creativity. The time away from a particular task need not last for years, as in Twain's case. You can enhance your creativity simply by letting the mind wander at times, perhaps as you listen to music, or even sleep on the problem. As Srini Pillay has said, "The brain operates optimally when it toggles between focus and unfocus, allowing you to develop resilience, enhance creativity, and make better decisions too."[33]

Taking a walk, in particular, can provide a powerful creativity boost, perhaps because it helps to unfocus our minds. Throughout history, great thinkers such as Charles Darwin, William Wordsworth, and Ludwig Van Beethoven took long walks on a daily basis.[34] Writing about the English poet Wordsworth's frequent strolls and hikes, Charles De Quincey said, "Wordsworth must have traversed a distance of 175 to 180,000 English miles—a mode of exertion . . . to which he has been indebted for a life of unclouded happiness, and we for much of what is most excellent in his writings."[35]

Psychological Distance

Further evidence of the benefits of unfocusing comes from our growing understanding of the concept of psychological distance. Sometimes, we benefit by stepping away from a problem in ways that cause us to think in a more abstract fashion. Achieving some psychological distance proves to be a powerful mechanism for enhancing creativity. As it turns out, we can achieve useful detachment in many different ways—not simply by taking a break or going for a walk.

Social Distance

Imagine that you were faced with the following problem to solve, and that you were struggling to arrive at a solution. What could you do to stimulate your creative juices?

> A prisoner was attempting to escape from a tower. He found a rope in his cell that was half as long enough to permit him to reach the ground safely. He divided the rope in half, tied the two parts together, and escaped. How could he have done this?

Evan Polman and Kyle Emich gave this problem to 137 undergraduates. They asked one-half of the research subjects to imagine that they were trapped in the tower and had to find a way to escape. Forty-eight percent of these students discovered the solution. The scholars directed the other half of the study participants to imagine that someone else was trapped. These participants had to find a way to help that other person escape from the tower. Roughly two-thirds of these individuals generated a correct answer to the problem. Achieving some psychological distance by imagining that you were solving the problem on someone else's behalf proved to be very beneficial. Why? When we are too close to a problem, we think more concretely and focus on granular details. We might even attend to irrelevant aspects of the situation. When we achieve some psychological distance, we tend to think in a more abstract and decontextualized manner. This type of broader conceptual thinking helps us generate more creative solutions.[36]

Which way did you think about the prisoner in the tower problem? Did it stump you? Perhaps if you had imagined someone else trapped in the tower, you would have come up with the solution: The prisoner could divide the rope in two lengthwise and tie the two halves together. Then, he or she will have a sufficient length of rope to climb down from the tower.

What about if you imagined yourself as someone else? Could putting yourself in others' shoes enhance your creativity? Would it matter whose shoes we chose to occupy? Denis Dumas and Kevin Dunbar decided to find out. They asked 105 individuals to complete a uses-of-objects task often utilized as way to measure divergent thinking capability. In this test, people must generate multiple original uses for well-known objects such as a fork or a hammer. Judges evaluated the responses for fluency (how many ideas did they generate?) and originality (how distinct were the ideas?).

Dumas and Dunbar asked one-half of the students to imagine themselves as "eccentric poets," and the other half to think of themselves as "rigid librarians." They chose these two roles because students tend to believe in these stereotypes. After receiving their role assignment, the students performed the uses-of-objects test for five items over the course of 10 minutes. The scholars then switched the stereotypes. The poets now imagined themselves as librarians and vice versa. The students spent ten more minutes completing the uses-of-objects test for five additional items. What did the results demonstrate? Students generated more ideas when they imagined themselves as eccentric poets as opposed to rigid librarians (see Figure 6.1)! Moreover, the poets demonstrated further originality as well![37]

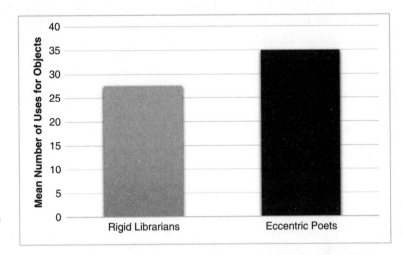

Figure 6.1   Do Eccentric Poets Generate More Ideas Than Rigid Librarians?

Source: Data drawn from Dumas and Dunbar, "The Creative Stereotype Effect"[38]

Leaders can create social distance for their team members as they organize meetings and design decision processes. Take the case of Kevin Dougherty, an executive at Sun Life Financial. As e-commerce exploded in different industries, he wanted his team members to consider the disruptive forces that might affect the insurance business in the years ahead. He worried that the firm's traditional strategic planning process might not be the best mechanism for generating creative new business proposals. The process generally favored incremental improvements to the existing business.[39]

Dougherty and his director of strategic planning, Thomas Hollman, decided to plan an off-site retreat for the management team of their division. They divided the team into four subgroups, each with four members. They asked each subgroup to develop proposals for new web-based ventures that Sun Life could launch. Dougherty and Hollman assigned each individual to play a specific role in their subgroup: chief executive officer (CEO), chief operating officer (COO), chief financial officer (CFO), and chief marketing officer (CMO). They directed each person to consider the interests and perspective of the role to which they had been assigned.

During the retreat, Dougherty and Hollman assigned people to play a role other than the one that they typically occupied in the organization. For instance, a sales executive stepped into the role of the CFO, and finance managers adopted the perspective of operations leaders. Team member Janice Wallace described how people embraced these roles enthusiastically during the off-site retreat: "They were so into their roles, I couldn't believe it. I think the different roles helped them to get out of their normal way of thinking as well and maybe feel a little safer when they presented new ideas."[40] Client services executive Rebecca Johnston described the value of stepping into others' shoes: "I think taking on different roles really helped us to see that, if we're going into e-business, which is a whole new business for us, we also have to be thinking more broadly. We can't just be in our own little silos."[41] Marketing executive Brigitte Parent noted that the process helped people take the blinders off and stimulated divergent thinking: "When you develop a strategy, you want everyone to be convinced that it is the right way to go, and to move in the same direction. One of the dangers, though, is that you may be blind-sided... This was a great meeting because it helped us ask ourselves, 'Are we missing something? Is there something that could surprise us?'"[42]

Design thinkers often create social distance during a project by imagining themselves as users and trying to walk a mile in their shoes. They sometimes role-play how different types of people may experience the purchase and consumption of a good or service. These empathetic efforts stimulate creative thinking. For instance, IDEO's Kristian Simsarian decided to imagine himself as a patient when designing a new hospital wing for the SSM DePaul Health Center in Saint Louis. He did more than imagine though. He actually pretended to have a foot injury and went to the emergency room at the hospital. Simsarian experienced the check-in process, the completion of paperwork, and the insurance verification procedures, a set of activities that often befuddle patients. He endured the lengthy waiting times at various points during his stay. He experienced the frustrations, anxiety, and confusion that many patients feel when they seek medical treatment. In so doing, Simsarian developed empathy for patients, and he began to discover their latent needs, those wants and desires that are sometimes difficult to articulate.

Eventually, Simsarian and his team generated a series of original ideas for how to enhance the patient experience.[43]

Leaders also can assign members of their management team to role-play key competitors as a way to spark divergent thinking during strategy meetings. In the military, war gaming often involves a red team competing against a blue team in a training exercise. The blue team plots strategy and tactics for the United States. The red team consists of a group of military personnel that have been directed to role-play the enemy. Highly effective red teams have devised bold and unorthodox strategies that blue teams did not anticipate. Asking people to step into the shoes of the enemy has sparked divergent thinking and caused military leaders to rethink their strategies and plans. Many corporate executives have embraced this concept of taking on the role of the competition, and it has spurred managers to question the conventional wisdom, challenge longstanding assumptions, and develop original ideas.[44]

Physical and Cultural Distance

> A dealer in antique coins got an offer to buy a beautiful bronze coin. The coin had an emperor's head on one side and the date 544 B.C. stamped on the other. The dealer examined the coin but instead of buying it, he called the police. Why?

Lile Jia and his colleagues gave this problem, and several similar puzzles, to 132 students. They gave the individuals two minutes to solve each problem. Jia's team of scholars claimed to be collecting data on behalf of another research institution. They told some students that the other university was 2,000 miles away, while telling others that it was just two miles from their current location. The participants in the control group received no information at all about the proximity of the university collecting the data in this study. Amazingly, students who thought they were solving the problem for a university 2,000 miles away were much more likely to arrive at a correct solution than the others in the study. Once again, creating some psychological distance, even in such a seemingly trivial manner, had a substantial impact on creative thinking.[45]

How can we take advantage of these beneficial effects of creating physical distance? Leaders can encourage their people to travel and experience other cultures. Saint Augustine once said, "The world is a book, and those who do not travel read only a page."[46] Travel disrupts our normal routines, and novelty stimulates the brain. Living in another nation can open people's eyes to new perspectives and cause people to question the typical ways that things are done in their home country. Immersing yourself in another culture provides insight as to how people in other countries work, live, and play—and perhaps most importantly, how they approach certain types of problems. David and Tom Kelley of IDEO have argued that travel awakens our minds and causes us to notice things that we often take for granted. They explain:

> Things stand out because they're different, so we notice every detail, from street signs to mailboxes to how you pay at a restaurant. We learn a lot when we travel not because we are any smarter on the road, but because we pay such close attention. On a trip, we become our own version of Sherlock Holmes, intensely observing the environment around us. We are continuously trying to figure out a world that is foreign and new. Too often, we go through day-to-day life on cruise control, oblivious to huge swaths of our surroundings. To notice friction points—and therefore opportunities to do things better—it helps to see the world with fresh eyes.[47]

Fashion designers often seek inspiration from different parts of the world. How does travel shape their creativity? Frédéric Godart and his colleagues have studied the creative directors of 270 high-end fashion houses over roughly a decade. They gathered information about the life and work histories of these men and women. The researchers used ratings generated by the famous *Journal du Textile* to evaluate the creativity of fashion collections. The magazine compiles its ratings each season by asking dozens of buyers at major retailers to evaluate fashion collections. *Journal du Textile* also asks a team of expert journalists to judge the creativity of the work produced by designers. Godart and his colleagues discovered that creative directors achieved higher ratings if they had spent a considerable amount of time in other countries. In particular, immersing yourself in a few other countries proved most beneficial, far more effective than simply visiting many different places for brief periods of time.[48]

Blythe Harris, co-founder and creative director of the jewelry firm Stella & Dot, believes passionately in the virtues of travel. She takes several "inspiration trips" each year, immersing herself in the local culture in hopes of stimulating her brain. She says:

> A favorite recent trip was to Japan, where everything was different: sushi for breakfast, innovative street style, and an inspired service culture. When you are experiencing something totally new, your mind slows down and your imagination really ignites. I also go to Europe twice a year and visit vintage dealers in Paris and London. I love pieces with a history, and I find vintage to be a great starting point.[49]

The Beatles took an inspirational trip of their own in February 1968. They went to study transcendental meditation with the Maharishi Mahesh Yogi in Rishikesh, India. John Lennon and George Harrison spent two months living and studying there, while their bandmates visited for a shorter time. The musicians ate vegetarian meals together, learned about local culture, and embraced Indian attire. Journalist Lewis Lapham reported, "Like the other Beatles, Harrison delighted in the costumes—embroidered overblouses, fanciful brass pendants, cotton pajama trousers broadly striped in bright colors, robes for all occasions."[50] The group even learned about local animal life, as a series of creatures made their way into the area where they were staying. Mike Love of The Beach Boys visited while the Beatles were there, and he remembers the wildlife vividly: "Spiders, stray dogs, and even an occasional tiger roamed the grounds. The night sounds were a shrill chorus of wildlife—peacocks, crows, and parrots. The wails and cackles may have unnerved some, but I felt at peace."[51]

The group did not intend to work on an album while on the trip. Harrison and Lennon, in particular, aimed to focus on their spiritual life above all else. McCartney had other thoughts. He carried his guitar around with him often. McCartney recalls that Harrison balked at discussing plans for future recordings: "I remember talking about the next album and he would say, 'We're not here to talk about music—we're here to meditate.' Oh yeah, all right, Georgie Boy. Calm down. Sense of humor needed here, you know. In fact, I loved it there."[52]

As it turns out, the Beatles wrote prolifically while on the trip despite Harrison's desire to prioritize spirituality before music. Lennon recalled, "We wrote about thirty new songs between us. Paul must have done about a dozen. George says he's got six, and I wrote fifteen."[53] The group's *White Album* features many tunes originally conceived in India, and the *Abbey Road* album also includes two songs written during the trip. Some of the Beatles' best work emerged during and soon after this immersion experience in Rishikesh, India.[54] Cultural and physical distance indeed rouses the brain.

Before we move on, I have to ask. Have you solved the problem about the bronze coin yet? Consider the meaning of the term "B.C." In the year 544 B.C., no one knew about the birth of Jesus Christ in Bethlehem. It would not happen for another five and a half centuries. Therefore, a coin at that time could not be marked "B.C." It was indeed a fraud!

Temporal Distance

Imagine you were asked to time travel, to imagine yourself in the distant future. Would that have an impact on your creativity? Jens Förster and his colleagues set out to determine the answer to that question. They gave 138 individuals a creative task to complete. Before giving them the task, the researchers asked some of the study participants to imagine themselves one year in the future, while directing others to think about what tomorrow would be like. A final set of participants did not receive any instructions regarding time travel. The people thinking about the distant future developed the most creative solutions in this exercise, as judged by outside experts (see Figure 6.2). Importantly, the study demonstrated that the creativity boost occurred when people worked on abstract idea-generation tasks, but not when individuals solved concrete problems. Temporal distance inspired creativity, similar to the effects

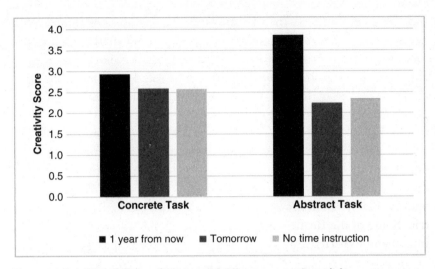

Figure 6.2   The Effect of Temporal Distance on Creativity

Source: Data drawn from Jens Förster, Ronald Friedman, and Nira Liberman, "Temporal construal effects on abstract and concrete thinking: consequences for insight and creative cognition"[56]

produced by achieving social, physical, and cultural distance. Time travel worked because it encouraged abstract, high-level, conceptual thinking.[55]

Jeff Bezos and the people at Amazon use "time travel" frequently to stimulate innovative thinking about new products and services. Andy Jassy, Senior Vice President of Amazon Web Services, explains that developers in his organization do not begin writing software code for a new project until they have drafted a hypothetical press release for their new product offering. People use this same process throughout the firm. Of course, the company will not actually publish press releases regarding these initiatives for months or even years, as it may take that long to bring the innovation to market.[57]

Amazon's Ian McAllister describes this process as "working backwards." Individuals must step into the future and imagine the moment when their new product or service hits the market. Writing the press release helps them to envision that future scenario. Managers must imagine the product features that will be stressed the most, the value proposition that they will tout, and the customers they will choose to target. They have to consider how customers will receive the news presented in this press release. Will it delight Amazon's users? What type of questions or concerns will customers have? After this time-travel experience, employees work backwards and engage in the creative problem solving that will be required to deliver a hit product in the future.[58]

Table 6.1 describes techniques that individuals and teams may use to achieve various forms of psychological distance.

Table 6.1 Strategies for Achieving Psychological Distance

Type of Distance	Concept	Strategies
Social Distance	Creativity increases if you: Imagine that someone else is facing your problem or situation. Imagine yourself as some other person.	Role-play competitors through a red teaming exercise. Ask executives to step into the shoes of leaders in other roles within your firm. Conduct a "walk a mile" exercise so as to empathize with your customers.
Physical and Cultural Distance	Creativity increases if you: Imagine that are far away from your current location. Travel to and live in different cultures.	Include people on your team who have lived in other countries. Provide international opportunities and assignments for your people. Take "inspiration trips" focused on learning about key aspects of other cultures.
Temporal Distance	Creativity increases if you: Imagine that you are making a decision in the distant future.	Think about communicating your concept to customers in the future and work backward to shape your idea. Assign people to create scenarios for how the industry will evolve in the next few years, and to imagine working in that situation.

Focus + Distance

Creative breakthroughs certainly occurred at Slane Castle in 1984. However, U2's *The Unforgettable Fire* did not emerge complete when the band members left the grounds of this historic site. A very brief break played a key role in their creative process. Bono and his fellow bandmembers gathered later that summer at Windmill Lane Studios in Dublin, and they continued to revise the album. They struggled at times to finish their songs, continuing to tinker for weeks on end. The band members worked feverishly toward the end of their time there, recording day and night. Bono noted, "I never finished great songs like 'Bad.' Classics like 'Pride (In The Name Of Love)' are left as simple sketches."[59] Whatever his feelings about these songs, U2 released *The Unforgettable Fire* on October 1, just three months after leaving Slane Castle. As Steve Jobs said, "Real artists ship."

The best creative thinkers toggle between focus and unfocus throughout their lives, much like Twain, Tolkien, and Wordsworth. They hole themselves up in a castle or war room, yet pigeonhole a manuscript and head off on a steamboat trip when they hit a roadblock. Visionary thinkers work feverishly at times with single-minded attention to a particular project. They seek seclusion and avoid the perils of constant multitasking and frequent interruptions. However, these trendsetters and leading-edge thinkers also take purposeful breaks and strive for psychological distance to stimulate their creativity. They know when to step away. As Twain once wrote, "I made the great discovery that when the tank runs dry you've only to leave it alone and it will fill up again in time, while you are asleep—also while you are at work at other things and are quite unaware that this unconscious and profitable cerebration is going on."[60]

Endnotes

1. "Bono Speaks," *U2 Magazine 10*, February 1, 1984 (www.atu2.com/news/bono-speaks.html, accessed January 24, 2018).

2. "War," U2.com (www.u2.com/music/Albums/4004/War, accessed January 24, 2018).

3. Steve Pond, "U2: The Joshua Tree," *Rolling Stone*, April 9, 1987 (www
   .rollingstone.com/music/ albumreviews/the-joshua-tree-19870409, accessed
   January 24, 2018).

4. Lyricshall, "The Unforgettable Fire," Lyricshall.com, n.d. (www.lyricshall.com/
   albums/U2/The+Unforgettable+Fire/, accessed January 24, 2018).

5. Joshua Klein, "Brian Eno and Daniel Lanois Remember the Making of U@'s
   Unforgettable Fire," Pitchfork.com, October 23, 2009 (pitchfork.com/news/
   36883-brian-eno-and-daniel-lanois-remember-the-making-of-u2s-
   unforgettable-fire/, accessed January 29, 2018).

6. Alex Dobuzinskis, "U2 Took "Unforgettable" Trip to Castle for '84 Album,"
   Reuters.com, October 26, 2009 (www.reuters.com/article/us-lanois/u2-took-
   unforgettable-trip-to-castle-for-84-album-idUSTRE59P56220091026,
   accessed January 24, 2018).

7. *The Making of the Unforgettable Fire*, directed by Barry Devlin, Island Records,
   1984.

8. "The Unforgettable Fire," U2.com, n.d. (www.u2.com/music/Albums/4006/
   The+Unforgettable+Fire, accessed January 24, 2018).

9. MessyNessy, "Then & Now: The Rolling Stones' French Villa of Debauchery,"
   MessyNessyChic.com, December 23, 2014 (www.messynessychic.com/2014/
   12/23/then-and-now-the-rolling-stones-french-villa-of-debauchery/, accessed
   January 25, 2018).

10. "The Untold Story of Bob Dylan's 'Basement Tapes'": Inside the New Issue,
    Rolling Stone, November 5, 2014 (www.rollingstone.com/music/news/the-
    untold-story-of-bob-dylans-basement-tapes-inside-the-new-issue-20141105,
    accessed January 25, 2018).

11. Trace William Cowen, "Kanye West Reportedly 'Holed Up' at the Top of a
    Mountain in Wyoming Working on New Album," Complex.com, May 9, 2017
    (www.complex.com/music/2017/05/kanye-west-holed-up-top-mountain-
    wyoming-working-new-album, accessed January 25, 2018).

12. CreativeHuddle, "Creativity: Better Alone or in Groups?" CreativeHuddle.co.uk
    (www.creativehuddle.co.uk/creativity-better-alone-or-in-groups, accessed
    January 22, 2018).

13. Thomas Oppong, "The Remarkable Science of Silence: How Solitude Enriches
    Creative Work," Medium, April 25, 2017 (medium.com/personal-growth/the-
    remarkable-science-of-silence-how-solitude-enriches-creative-work-
    be0eb19b15d0, accessed January 22, 2018).

14. This section benefits from my visit to Google in November 2017, during which
    time I met Jake Knapp and heard him present on the origins of the design sprint
    methodology. I am grateful to Kai Haley for inviting me to the inaugural Google
    Design Sprint conference, where I had the opportunity to speak with and learn

from Jake Knapp, as well as design sprint professionals from many firms including Airbnb, Facebook, Uber, Intuit, Dropbox, IDEO, and Microsoft.

15. Jake Knapp, John Zeratsky, and Braden Kowitz. *Sprint: How to Solve Big Problems and Test New Ideas in Just Five Days* (New York: Simon & Schuster, 2016), 38.

16. Jake Knapp, "Tour of the GV Sprint Room in San Francisco," LinkedIn, February 1, 2016 (www.linkedin.com/pulse/tour-gv-sprint-room-san-francisco-jake-knapp/, accessed January 29, 2018).

17. Gloria Mark, Victor Gonzalez, and Justin Harris. "No Task Left Behind? Examining the Nature of Fragmented Work," *Proceedings of the SIGCHI conference on Human factors in computing systems*, ACM, 2005, 321–330.

18. Kermit Pattison, "Worker, Interrupted: The Cost of Task Switching," *Fast Company*, July 28, 2008 (www.fastcompany.com/944128/worker-interrupted-cost-task-switching, accessed January 29, 2018).

19. Cyrus Foroughi, Nicole Werner, Erik Nelson, and Deborah Boehm-Davis, "Do Interruptions Affect Quality of Work?" *Human Factors*, 56(7), 2014, 1262–1271.

20. Eyal Ophir, Clifford Nass, and Anthony D. Wagner, "Cognitive control in media multitaskers," *Proceedings of the National Academy of Sciences*, 106(37), 2009, 15583–15587.

21. David Sanbonmatsu, David Strayer, Nathan Medeiros-Ward, and Jason Watson, "Who Multi-tasks and Why? Multi-tasking Ability, Perceived Multi-tasking Ability, Impulsivity, and Sensation Seeking," *PloS one* 8(1), 2013, e54402.

22. David Doughan, "J. R. R. Tolkien: A Biographical Sketch," The Tolkien Society, 2018 (www.tolkiensociety.org/author/biography/, accessed January 30, 2018).

23. "C. S. Lewis, J. R. R. Tolkien, and the Inklings," CSLewis.com, April 16, 2009 (www.cslewis.com/c-s-lewis-j-r-r-tolkien-and-the-inklings/, accessed January 30, 2018).

24. Maev Kennedy, "From the North Pole to Middle-Earth: Tolkien's Christmas Letters to His Children," *The Guardian*, December 19, 2017 (www.theguardian .com/books/2017/dec/19/north-pole-middle-earth-tolkien-christmas-letters-children, accessed January 30, 2018).

25. "JRR Tolkien Trained as a British Spy," *The Telegraph*, September 16, 2009 (www.telegraph.co.uk/news/uknews/6197169/JRR-Tolkien-trained-as-British-spy.html, accessed January 30, 2018).

26. "Best-Selling Novels the Critics Hated," Forbes.com, n.d. (www.forbes.com/ pictures/mfg45hihf/lord-of-the-rings-jrr-tolkien/#3710457d7b86, accessed January 30, 2018).

27. Ron Powers, *Mark Twain: A Life* (New York: Free Press, 2005).

28. Letter may be found at the Mark Twain Project Online (www.marktwainproject .org/, accessed January 31, 2018).

29. Henry Nash Smith, *Mark Twain: A Collection of Critical Essays* (Englewood Cliffs, NJ: Prentice Hall, 1963), 89.

30. "Mississippi River Tour 1882 Itinerary," TwainQuotes.com, www.twainquotes .com/Steamboats/Itinerary1882.html, accessed February 2, 2018).

31. Robert B. Brown, "One Hundred Years of Huck Finn," *American Heritage* magazine 35(4), June/July 1984 (www.americanheritage.com/content/one-hundred-years-huck-finn, accessed January 31, 2018).

32. Walter Blair, *Essays on American Humor: Blair through the Ages*, ed. Hamlin Hill (Madison, WI: University of Wisconsin Press, 1993), 199.

33. Srini Pillay, "Your Brain Can Only Take So Much Focus," *Harvard Business Review* (Digital Article), May 12, 2017.

34. Ferris Jabr, "Why Walking Helps Us Think," *The New Yorker*, September 3, 2014.

35. Rebecca Solnit, *Wanderlust: A History of Walking* (New York: Penguin Books, 2000), 77.

36. Evan Polman and Kyle Emich, "Decisions for Others Are More Creative Than Decisions for the Self," *Personality and Social Psychology Bulletin*, 37(4), 2011, 492–501.

37. Denis Dumas and Kevin Dunbar, "The Creative Stereotype Effect," *PloS one*, 11(2), 2016, e0142567.

38. Ibid.

39. This section draws from research that I conducted at Sun Life Canada and published in a Harvard Business School case study. See Michael Roberto, "Strategic Planning at Sun Life," Harvard Business School Case Study 9-301-084, August 1, 2001.

40. Roberto, "Strategic Planning at Sun Life," 13.

41. Roberto, "Strategic Planning at Sun Life," 13.

42. Roberto, "Strategic Planning at Sun Life," 15.

43. Tim Brown, *Change by Design: How Design Thinking Transforms Organizations and Inspires Innovation*, (New York: Harper Business, 2009).

44. I have learned about red teaming during visits to the U.S. Special Operations Command at Fort Bragg, Air War College, Naval War College, and U.S. Military Academy at West Point.

45. Lile Jia, Edward Hirt, and Samuel Karpen, "Lessons from a Faraway Land: The Effect of Spatial Distance on Creative Cognition," *Journal of Experimental Social Psychology*, 45(5) 2009, 1127–1131.

46. Father Mark Coiro, the pastor at my parish, loves this quote from St. Augustine, and I learned it from him.

47. Tom Kelley and David Kelley, *Creative Confidence: Unleashing the Creative Potential Within Us All* (New York: Crown Business, 2013), 77.

48. Frédéric Godart, William Maddux, Andrew Shipilov, and Adam Galinsky, "Fashion with a Foreign Flair: Professional Experiences Abroad Facilitate the Creative Innovations of Organizations," *Academy of Management Journal,* 58(1), 2015, 195–220.

49. Blair Farris, "On the Go with Blythe Harris," Peachy, August 2, 2017 (peachythemagazine.com/2017/08/on-the-go-with-blythe-harris/, accessed February 2, 2018).

50. David Chiu, "The Beatles in India: 16 Things You Didn't Know," *Rolling Stone,* February 12, 2018 (www.rollingstone.com/music/lists/the-beatles-in-india-16-things-you-didnt-know-w516195, accessed March 4, 2018).

51. Ibid.

52. Ibid.

53. The Beatles, *The Beatles Anthology* (San Francisco: Chronicle Books, 2000), 305.

54. Dave Swanson, "50 Years Ago: The Beatles Meet the Maharishi," UltimateClassicRock.com, February 15, 2016 (ultimateclassicrock.com/the-beatles-india-maharishi/, accessed March 4, 2018).

55. Jens Förster, Ronald Friedman, and Nira Liberman, "Temporal Construal Effects on Abstract and Concrete Thinking: Consequences for Insight and Creative Cognition," *Journal of Personality and Social Psychology,* 87(2), 2004, 177–189.

56. Ibid.

57. Jillian D'Onfro, "Why Amazon Forces Its Developers to Write Press Releases," *Business Insider,* March 12, 2015 (www.businessinsider.com/heres-the-surprising-way-amazon-decides-what-new-enterprise-products-to-work-on-next-2015-3, accessed February 5, 2018).

58. Andre Faria, "Try an Internal Press Release before Starting New Products," March 21, 2014 (medium.com/bluesoft-labs/try-an-internal-press-release-before-starting-new-products-867703682934, accessed February 5, 2018).

59. U2 and Neil McCormick, *U2 by U2* (New York: Harper Collins, 2009), 188.

60. PBS, "Mark Twain: About the Film," n.d., (www.pbs.org/marktwain/scrapbook/06_connecticut_yankee/page2.html, accessed January 31, 2018).

CHAPTER 7

The Naysayer Mindset

Every human being is entitled to courtesy and consideration. Constructive criticism is not only to be expected but sought.
—Margaret Chase Smith, U.S. Representative and
U.S. Senator from Maine (1940–1973)

On April 6, 1973, the Egyptian military attacked Israeli military forces in the Sinai Peninsula, and five divisions of Syrian troops launched a simultaneous offensive in the Golan Heights. The Israelis found themselves badly outnumbered and unprepared for war. The outbreak of armed conflict caught the Israeli Defence Forces by surprise. Though numerous warning signs had emerged in the weeks leading up to the war, top Israeli officials had downplayed the threat repeatedly.

Israeli military intelligence officials held strong beliefs about the possibility of war with the nation's neighbors in the Middle East. Former Israeli intelligence official Ephraim Kam reflected that, "The Israeli error began with a basic concept that the Arabs would not attack during the next two to three years, and every new development was adapted to this concept. In some cases, even a clear indication of imminent attack was not enough to change the established view."[1] For years, the Israeli military presumed that Egypt would not launch an attack unless it had developed air superiority over Israeli forces. A key intelligence source indicated that Egyptian Prime Minister Anwar Sadat believed that he needed to acquire long-range fighter-bombers from the Soviet Union to achieve the necessary airpower to combat Israel. Since the Soviets had not provided those aircraft by April 1973, Israeli officials remained steadfast that war was not imminent.[2]

At the end of 1972, Israeli intelligence officials reported that, "The probability that Egypt will try to cross the canal is close to zero."[3] As 1973 progressed, numerous signals of a possible attack emerged. Brookings Institution senior fellow Bruce Riedel concluded that, "The intelligence community adhered to its concept and interpreted the data collected to fit inside the box. The policy consumer of the intelligence estimate did not challenge the analysis, but rather reinforced it."[4] A few senior military officers tried to question the conventional wisdom. Major General Eli Zeira, director of military intelligence, rebuffed their concerns. Some experts have noted that Zeira did not welcome dissenting views.[5]

In the aftermath of the conflict, Prime Minister Golda Meir established a commission of inquiry led by Shimon Agranat, Chief Justice of the Israeli Supreme Court. The investigative board concluded that a massive intelligence failure had occurred, and it recommended numerous changes to prevent future surprises.[6] Following the publication of the report, the Directorate of Military Intelligence (known as AMAN) created a devil's advocate office. Former Israeli intelligence official Yossi Kuperwasser explains the role of this group:

> The devil's advocate office ensures that AMAN's intelligence assessments are creative and do not fall prey to groupthink. The office regularly criticizes products coming from the analysis and production divisions, and writes opinion papers that counter these departments' assessments. The staff in the devil's advocate office is made up of extremely experienced and talented officers who are known to have a creative, "outside the box" way of thinking.[7]

The devil's advocate office surfaces and challenges key assumptions embedded in the official intelligence estimates. The office also generates worst-case scenarios that should be considered.

The Devil's Advocate

Many business leaders have embraced the devil's advocacy technique over the years. While some research supports the notion that authentic dissenters are more effective than appointed contrarians, executives do not always have the luxury of waiting for opposing views to surface on their own.[8] Therefore, some leaders institutionalize critique, in hopes of enhancing the level of creative and critical thinking within their teams.

Kimball Hall, a senior executive at Genentech (a subsidiary of Roche), asks members of her management team to play different roles at times, so as to spark divergent thinking. One role focuses on finding the weaknesses and flaws in the primary proposal on the table.[9] Bob Pittman, founder of MTV and long-time chief executive in the entertainment industry, asks people, "What did the dissenter say?" when they come to him with a proposal. Sometimes, they tell him that widespread support exists for their recommendation. Pittman challenges them to find someone with an alternative point of view so that he can consider the contrarian perspective before making a final decision.[10]

Ori Hadomi, CEO of Mazor Robotics, discovered that his team used to fall into the overconfidence trap and seemed to generate rosy scenarios about the future on a consistent basis. He decided to remedy the situation:

> One of our takeaways . . . was to appoint one of the executive members as a devil's advocate. He's actually very challenging and he knows how to ask the right questions. He really makes sure to say to me, "Let's be more humble with our assumptions."[11]

Some leading designers have embraced devil's advocacy as a part of their creative process as well. At Google, Jake Knapp seeks to establish a collegial atmosphere where group members share ideas freely. However, he makes sure that each design sprint team includes a "troublemaker," a person likely to present contrarian perspectives. He explains that you might be hesitant about bringing this person into the process for fear that they will be disruptive. However, you have to learn to accept a bit of discomfort because contrarians provide a valuable contribution to the team. As Knapp explains:

> Troublemakers see problems differently than everyone else. Their crazy idea about solving the problem might just be right. And even if it's wrong, the presence of a dissenting view will push everyone else to do better work.[12]

Designers Jacklyn Burgan and Will Evans have created a structured way to introduce devil's advocacy into their design sprints. They call it a "ritual dissent" exercise. As someone presents his or her team's storyboard to others, the audience takes notes while remaining silent. After the brief presentation, people begin to critique the concept. While they provide feedback, the presenter turns his or her back. He or she must not respond to the comments at that time; the individual should simply record the feedback. By remaining silent, the presenter can listen carefully and not become preoccupied with formulating an immediate response to each remark. Burgan and Evans explain that the presenter must bring a "thick skin" to this exercise, while the others must be prepared to offer honest criticism in a constructive manner.[13]

Experimental research confirms the efficacy of devil's advocacy. David Schweiger and his colleagues conducted several studies comparing a consensus decision-making approach with a process in which some team members play the role of devil's advocate. In the consensus approach, team members worked together in a free and open discussion. Everyone had an opportunity to share their ideas. The group tried to arrive at an agreement in a collegial fashion. In the devil's advocacy method, each team divided itself into two subgroups. One subgroup developed a proposal to address the problem it was trying to solve, while the other subgroup critiqued the proposal. Each team tried to arrive at a decision through this iterative point-counterpoint process. Expert judges rated the quality of the decisions made using the devil's advocacy technique to be significantly higher than those made by teams utilizing the consensus method. Why? The higher-performing teams surfaced and probed assumptions more carefully, uncovered hidden risks, and engaged in more thorough evaluation of alternatives.[14]

My research suggests that the presence of a devil's advocate also improves problem solving by helping teams overcome shared information bias. For years, we have known that groups tend to spend a great deal of time discussing information shared by all members. Unfortunately, teams spend an insufficient amount of time uncovering, discussing, and integrating unshared or unique information (i.e., data that one person possesses, but others do not).[15] Of course, we bring people together on a team because we want them to pool their diverse expertise so as to develop solutions that no individual could derive on his or her own. The shared information bias means that we often do not achieve this desired synergy within our teams.

Brian Waddell, Sukki Yoon, and I decided to test whether the presence of a devil's advocate might help teams overcome shared information bias. We asked groups of students to solve a murder mystery developed by psychologist Garold Stasser, the researcher who discovered shared information bias. The mystery involves three suspects. Each team must sift through many clues to determine the murderer and exonerate the others. We gave some clues to all team members, while we distributed other clues to specific individuals. For some groups, we assigned one team member to play the role of devil's advocate. For other teams, we simply asked them to work collaboratively to solve the mystery. The groups with a devil's advocate enjoyed a higher success rate in this exercise. The presence of someone offering a contrarian point of view facilitated the sharing of data and the integration of distributed knowledge.[16]

Creativity Killer?

Some people do not have a fond view of the devil's advocate. They believe that tenacious critics who always look to poke holes in new ideas can stifle creativity. IDEO's Tom Kelley describes these people as "idea wreckers" who dwell on the negative far too often. He pulls no punches in providing his perspective on this technique:

> "Let me just play devil's advocate for a minute . . . " Having invoked the awesome protective power of that seemingly innocuous phrase, the speaker now feels entirely free to take potshots at your idea and does so with impunity . . . Devil's advocates remove themselves from the equation and sidestep individual responsibility for the verbal attack. But before they're done, they've torched your fledgling concept . . . The devil's advocate may be the biggest innovation killer in America today. What makes this negative persona so dangerous is that it is such a subtle threat. Every day, thousands of great new ideas, concepts, and plans are nipped in the bud by devil's advocates.[17]

Kelley's assessment cannot be dismissed readily, because it reflects negative experiences many of us have had. We have become frustrated at times with those who always look on the dark side without offering alternative solutions. No one enjoys hearing from a know-it-all who has a rebuttal for every idea we propose. We have seen how devil's advocates induce self-censorship on the part of other team members. Our creative output has suffered because some colleagues withhold creative ideas if they witness others' proposals being decimated by a merciless critic. After a while, many individuals have learned to tune out their team's "Chicken Little" who always screams that the sky is falling. In some cases, the devil's advocate puts others on the defensive and incites interpersonal conflict. Personality clashes and personal friction fracture the team and impede its ability to generate and implement ideas moving forward.

Unfortunately, in many organizations today, the critics and naysayers outnumber the idea generators and the doers by a wide margin. In some sense, we have become a society poised to criticize at every opportunity. We review products on Amazon, restaurants on Yelp, and drivers on Uber. If we have a negative experience, we tell our friends on Facebook and our followers on Twitter. In past years, we listened to Gene Siskel and Roger Ebert evaluate films. Today, we all have become movie critics through sites such as Rotten Tomatoes. The democratization of critique has served us well in many ways, but perhaps the naysayer mindset has become an "innovation killer" in some organizations.

In school, we hear faculty members constantly talk about the need to teach critical thinking. Too often, though, it means that our students spend far more time analyzing what others have done, rather than generating new ideas and discoveries. In English class, we critique an author's plot devices and character development. In history class, we second-guess the decisions made by presidents, prime ministers, and military leaders. In business school, we analyze the flawed strategic choices made by the chief executive in each case study. Too often, we (and that includes me) reward the person who acts like the smartest person in the room by always finding fault and poking holes in others' ideas. We enable those who choose the easy path by offering criticism without proposing new solutions.

We should teach our students to evaluate thoughtfully and critically. However, we need them to create works of art, devise concrete policies to solve social problems, and develop their own business plans. We have to cultivate a generation of doers, not just "Monday-morning quarterbacks." Perhaps we need to remind ourselves of the wise words once shared by President Theodore Roosevelt:

> It is not the critic who counts; not the man who points out how the strong man stumbles, or where the doer of deeds could have done them better. The credit belongs to the man who is actually in the arena, whose face is marred by dust and sweat and blood; who strives valiantly; who errs, who comes short again and again, because there is no effort without error and shortcoming; but who does actually strive to do the deeds; who knows great enthusiasms, the great devotions; who spends himself in a worthy cause.[18]

Should organizations abolish the devil's advocate then? Of course not. New ideas flourish in the presence of some creative tension. Lennon and McCartney did not always see eye to eye, and their songwriting benefited from the clash of perspectives. Candid critique during the filmmaking process helps Pixar produce movies that audiences love. We need people on our team to challenge new ideas and to stimulate dissent and debate. However, we need the right kind of devil's advocates. We can't let the naysayer mindset overwhelm our organizations and stifle new ideas at every turn. We need devil's advocates who contribute to the dialogue in a constructive fashion, who help supercharge our creativity rather than squelching it. To do that, we must think carefully about the *who, when,* and *how* of devil's advocacy.

Who Plays the Devil's Advocate?

Born in 1914, Henry Wallich came of age during the Weimar Republic in Germany. He experienced the hyperinflation of the 1920s, and it left a mark upon him. Years later, he would vividly remember paying billions of German marks as the price of admission to a local swimming pool when he was nine years old.[19] When Wallich grew up, he moved to the United States, earned his doctorate in economics from Harvard, and became a faculty member at Yale. In 1974, President Richard

Nixon appointed him as a member of the Federal Reserve Board of Governors.

During the 1970s, inflation rose significantly in the United States. In Wallich's first year at the Federal Reserve, the consumer price index rose by 11.04 percent. While inflation eased a bit in the next few years, it increased again later in the decade, peaking at 13.50 percent in 1980.[20] Based in part on the experiences of his youth, Wallich served as the most ardent inflation hawk on the Federal Reserve Board of Governors during this era. He once said, "I regard inflation as a form of fraud."[21] He felt that the government had an obligation not to cheat its own citizens by devaluing the currency.

Wallich cast more dissenting votes (27) than any member in the Federal Reserve's history. He repeatedly argued for more restrictive monetary policy, in hopes of taming inflationary pressures. Wallich cast eight dissenting votes in 1980 alone.[22] Steven Rattner once wrote about the inner workings of the Federal Open Market Committee for the *New York Times*. He explained that Wallich's views were "taken less seriously" because he dissented so often. Many members "quietly tolerated" Wallich when he stressed time and again the perils of growing the money supply too quickly.[23] Journalist William Greider wrote that Wallich appeared to be "stuck like a broken record – predictably repeating himself regardless of circumstances."[24] Wallich once reflected on the dynamic within the committee as he cast dissenting votes time after time: "It's not a pleasant thing to have to keep dissenting (he said in a sorrowful tone). It makes one quite useless to dissent so often... To be a constant dissenter is a fruitless thing."[25]

Unfortunately, it took years for the Federal Reserve to adopt the harsh measures required to reduce inflation. Meanwhile, the United States experienced a decade of ruinous increases in the price level. If you purchased an item in 1972 for $100, it would have cost $231, on average, a decade later.[26] Eventually, Chairman Paul Volcker, with the support of President Ronald Reagan, tightened monetary policy considerably and reduced the inflation rate to 3.2 percent in 1983, and it remained relatively low for years.[27]

What can we learn from Wallich's story? For most teams, the same person cannot play the devil's advocate during every meeting. That contrarian voice risks sounding like a broken record, no matter how knowledgeable and experienced that person is. Team members become numb to the constant drumbeat of negativity, and they tune that voice out. Hard-charging, never-satisfied athletic coaches can encounter this problem.

Take the case of former Chicago Bulls coach Tom Thibodeau. He took over a mediocre basketball team in 2010, one that had eked into the playoffs and lost in the first round. During Thibodeau's first year, the team won 62 games, tying for the most wins by a rookie coach. Thibodeau earned the NBA Coach of the Year Award.[28] The team played with incredible defensive intensity under the watchful eye of their demanding coach. Players described him as relentless in his pursuit of excellence. He scrutinized his team's play meticulously. Eventually, though, some players began to tune him out. Reflecting on the coach's criticism during preseason one year, star player Derrick Rose once said, "I'm numb to it now, used to it."[29] The Bulls fired Thibodeau in 2015. Journalist Steve Rosenbloom wrote:

> At some point, no matter their success, relentless coaches hit their shelf life, and it frequently happens sooner than you think and more dramatic in the way it plays out. Players just stop listening. All at once. Relentless coaches turn their speakers up to 11. No matter. Deaf ears.[30]

To avoid this broken-record phenomenon, leaders must consider rotating the role of devil's advocate. Kevin Lofton serves as CEO of Catholic Health Initiatives, one of the largest nonprofit healthcare systems in the United States. He designates a different member of his team to play the devil's advocate during each important decision-making process.[31] Corporate governance expert John Zinkin recommends rotating the role among board members to ensure productive dialogue during directors' meetings.[32]

Leaders also should consider appointing two devil's advocates rather than one. In many groups, the majority readily dismisses a lone dissenting voice. People can convince themselves that this individual must have arrived at an erroneous conclusion due to flawed reasoning or a lack of relevant expertise. It may be more difficult to discount or marginalize the contrarian viewpoints of two team members. Moreover, a dissenter faces tremendous pressures to conform in many groups. The devil's advocate may be able to resist the pressure to go along with the majority if he or she can find social support within the group, in the form of a second rebellious voice.

In a famous experiment, Solomon Asch once asked eight individuals to evaluate three lines of unequal lengths, and to select the one that matched the length of one line shown separately. The correct answer was unambiguous. As it turns out, seven of the eight people served as confederates; they had met with Asch earlier. He instructed the confederates to respond at times with incorrect judgments, and to do so unanimously and publicly. One person remained clueless to the manipulation – a minority of one. Asch wanted to see if that person would conform to the majority opinion, even though that view was clearly incorrect. Indeed, 74 percent of the individuals placed in the minority position conformed at least once during 12 trials; 30 percent went along with the majority's incorrect judgment in at least 6 of 12 trials![33]

Asch made a small adjustment in his next round of experiments. He gave the naïve research subject a partner during each trial. Asch directed this partner to respond before the naïve participant, and to always respond with the correct answer. Six confederates formed the majority, and they responded unanimously with the wrong solution. In this case, the naïve participants went along with the majority view much less often than in the first experiment. The social support of a partner reduced the pressures for conformity a great deal.[34]

Years later, social psychologist Serge Moscovici flipped this experiment on its head. He placed two confederates in a group with four naïve participants. Moscovici informed the two confederates that they should provide consistently incorrect responses during a color perception task. When shown a series of slides that were clearly different shades of blue, these two individuals responded that they thought the slides were green, per Moscovici's instruction. The confederates managed to persuade 32 percent of the naïve participants to respond incorrectly (answering green instead of blue) at least once during the experiment. A consistent minority opinion expressed by two individuals can sway the judgments of those who find themselves in the majority within a team.[35]

Does a minority of two exhibit more influence than a lone dissenter? To examine this question, Jack Arbuthnot and Marc Wayner asked groups to review lawsuits and come to a consensus regarding the amount the claimant should receive. The scholars placed either one or two confederates on each team. The confederates argued for a very different award amount than the majority. The minority of two proved to be much more influential than the lone dissenter. Two contrarian voices working together managed to shift the award amount considerably more than a single opposing voice.[36]

Taken together, these studies provide strong support for the recommendation that teams should assign two devil's advocates rather than one. An Aon Hewitt survey of corporate directors bolsters this contention. They administered a questionnaire to 120 British directors as well as 300 well-educated members of the general public. For both samples, Aon Hewitt reported that people were much more likely to take feedback and criticism seriously if espoused by two dissenters rather than one.[37]

Perhaps the most famous instance of devil's advocacy during a high-stakes decision provides further evidence of the power of two dissenting voices. As President John Kennedy grappled with the Cuban Missile Crisis in October 1962, he assigned two men, Theodore Sorensen and Robert Kennedy, to serve as intellectual watchdogs during the deliberations amongst his senior advisers. The president asked these two trusted confidantes to "relentlessly pursue every bone of contention in order to prevent errors arising from too superficial an analysis of the issues."[38] Secretary of Defense Robert McNamara noted that their probing questions helped the group make a wise decision that enabled the country to avert catastrophe.[39]

When Should We Play the Devil's Advocate?

Before becoming famous as a cast member on *Saturday Night Live*, Tina Fey took classes and later performed with the Second City improvisational comedy troupe based in Chicago. At this theater, Fey learned the "yes, and" principle of improv comedy. She explains the concept by referring back to a story she once heard about the great comedian Joan Rivers. Decades earlier, Rivers began an improv scene with her partner by exclaiming, "I want a divorce." Her spouse responded, "But what about the children?" Rivers shot back quickly, "We don't have any children!" Fey explains that the comment elicited a burst of laughter from the audience, but "it killed the scene."[40] Why? Rivers had not built upon what her partner had said. Fey explains:

> Obviously in real life you're not always going to agree with everything everyone says. But the Rule of Agreement reminds you to "respect what your partner has created" and to at least start from an open-minded place. Start with a YES and see where that takes you . . . . The second rule of improvisation is not only to say yes, but YES, AND. You are supposed to agree and then add something of your own.[41]

How does this improv comedy principle apply to the role of the devil's advocate in organizations? In the early stages of a decision-making process, we should consider withholding critique. At IDEO, they describe this practice as deferring judgment. You do not criticize proposals during the initial idea-generation phase. Why? People may engage in self-censorship if they watch others' ideas become dismantled quickly. You want to create a safe atmosphere where people do not fear having their ideas mocked or ridiculed. Moreover, we want people to feel comfortable putting forth solutions that might seem infeasible or impractical, perhaps even a bit crazy.

At Google, Jake Knapp does not advocate group brainstorming in the manner conducted at IDEO. He recommends that people generate ideas individually during the early phase of a design sprint. He calls it "work alone together."[42] He provides people time and space to generate ideas free of social influence and peer critique. Then they share their ideas with colleagues and begin to work collaboratively. My colleagues and I have combined the IDEO and Google approaches with much success in our work with students and executives. We provide time for people to work alone, and then we engage them in ideation exercises as a team. Nevertheless, we try to keep the devil's advocate at bay in the early stages of the problem-solving process.

If you cannot critique ideas during this initial problem-solving phase, how should you respond to others' proposals or concepts? You might propose an altogether different option. In addition, IDEO's designers advocate building on each other's ideas. In other words, augment what others have said and advance that line of thinking. Apply the "yes, and" principle. Unfortunately, in many organizations, people respond "yeah, but" to many original ideas. "Yeah, but we have tried that before in this organization." "Yeah, but we don't have the resources to implement that idea." "Yeah, but I don't think senior executives will support that recommendation." To stimulate creativity, we have to replace "yeah, but" with "yes, and" during the early phases of decision making (see Figure 7.1).

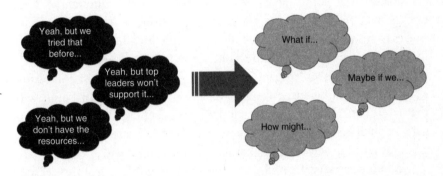

Figure 7.1 **"Yeah, but. . ." vs. "Yes, and. . ." Thinking**

The "yes, and" principle does not mean that we are eliminating the devil's advocate. Sometimes, people think that "yes, and" means eradicating conflict. Not at all! As innovation expert Scott Berkun has said, "The intention of brainstorming is not to eliminate critique, but simply to postpone it."[43] Once a team has generated a diverse array of options, individuals can begin to offer constructive critiques. Dissent and debate can improve those ideas and/or lead to the generation of new alternatives.

How Should We Play the Devil's Advocate?

How you express dissent matters a great deal. People play the devil's advocate in two distinct ways. Some devil's advocates strive to find fault with each proposal advanced by colleagues. They express their arguments with certainty, rather than acknowledging what they do not know or what remains ambiguous. They ask few questions; instead, they make definitive pronouncements. These dissenters tear apart proposals without offering alternative solutions. They examine only the downside risks, rather than viewing the benefits and costs of a particular recommendation in a balanced manner. They view a debate as a contest in which they aim to emerge victorious, perhaps at the cost of damaging interpersonal relationships. Deliberations become zero-sum games in which one party must win at another's expense. These counterproductive devil's advocates often repeat themselves stridently when others do not acknowledge the validity of their arguments. They put others on the defensive, provoke interpersonal conflict, and stop far too many good ideas in their tracks.

Fortunately, some devil's advocates behave quite differently. They engage in active listening, ask thoughtful questions, and open up people's minds to new possibilities. They do not simply list all the reasons why an idea won't succeed. Instead, they ask, "How might we make this work?" These dissenters seek to gain insight as to why others hold contrasting assumptions and beliefs. An effective line of inquiry might begin with the statement: "Help me understand why you believe . . . "[44] They admit gaps in their own understanding of a situation. Learning becomes the goal, rather than winning the battle of competing ideas.

When necessary, helpful dissenters encourage their team to reframe the problem. Perhaps, the group has defined the challenge too narrowly or too broadly. At LUMA Institute, design thinking experts offer a wonderful example of the power of reframing a problem. They ask you to imagine how Steve Jobs, Tony Fadell, Jony Ive, and others developed the iPod. They may have started out asking themselves, "How can we design a better MP3 player?" Fortunately, the team at Apple reframed the problem: "How might we create a more enjoyable music experience?" By casting the problem in a different light, the team did not focus on simply designing a terrific device, but instead created iTunes and delivered a holistic solution for consumers.[45]

Constructive devil's advocates seek to generate new options, rather than simply poking holes in the existing ideas. Perhaps people feel uncomfortable proposing an alternative that they consider unconventional. They make it safe for people to bring those ideas forward. An effective intervention might be, "Does anyone have an alternative that some might consider far-fetched or implausible?" They help a team make assumptions explicit and then probe them carefully. Rather than simply arguing that a supposition is invalid, these positive contributors ask: "What if that assumption proved to be false? How might that change our views and recommendations?" Finally, constructive devil's advocates encourage everyone to treat opinions as testable hypotheses, rather than proven facts. They ask, "How might we prove or disprove that theory? What new data might we gather or what experiment could we conduct?"

The best devil's advocates practice the Socratic method, rather than delivering a lecture. Consider a famous scene in the 1973 film *The Paper Chase*, set at Harvard Law School. On the first day of class, Professor Kingsfield calls on first-year student James Hart to open the discussion. The faculty member asks the student to discuss a medical malpractice case that everyone has been instructed to read in advance. Hart reacts in shock; he expected Kingsfield to deliver a lecture on the first day. After class, Hart races to the restroom and vomits. Welcome to law school!

Professor Kingsfield eventually explains his teaching philosophy to the first-year students. His approach centers on asking thought-provoking questions:

> We use the Socratic method here. I call on you, ask you a question, and you answer. Why don't I give you a lecture? Because through my questions, you learn to teach yourselves ... questioning and answering. At times, you may feel that you have found the correct answer. I assure you that this is a total delusion on your part. You will never find the correct, absolute, and final answer. In my classroom, there is always another question. We do brain surgery here ... You teach yourselves the law, but I train your minds. You come in here with a skull full of mush, and you leave thinking like a lawyer.[46]

An effective dissenter asks the right type of questions, which are shown in Table 7.1. Anthony Restivo, a designer at DraftKings, explains that he has been trained to ask open-ended questions when conducting customer research, rather than simple yes/no queries. Moreover, Restivo has learned to avoid posing leading questions, in hopes of hearing the answer he already has in mind. Effective contrarians behave much

Table 7.1   Asking the Right Questions

Type of Question	Sample Phrasing
Option Generation	What are some other options for solving this problem? Don't worry if it sounds foolish or infeasible, or whether it will prove financially viable...
Cause–Effect	Why might that action cause the outcome we desire? Help me understand the cause–effect relationship in more depth...
Assumption Probing	What must be true for this plan to succeed? Let's try to list our implicit assumptions and distinguish them clearly from the facts in this case...
Pre-mortem	What if execution does not go as planned? Let's imagine that we fail. What would that look like? Help me understand what could go wrong...
Role Play	What would a different set of people do if they were in our shoes? Let's try to determine why others might approach this problem in another manner...
Test and Learn	How might we design a test or experiment to determine if this plan of action will succeed? What could we learn from an experiment? Think about how we might test our ideas quickly and inexpensively...

like the best designers. They inquire in ways that invite free-flowing dialogue, and they avoid steering the discussion toward pre-ordained solutions.[47]

Constructive devil's advocates do not pretend to have all the answers. In fact, they acknowledge that they do not have superior knowledge about the subject at hand. They provoke dialogue and debate. They exhibit intellectual curiosity rather than certitude. They use questions to promote collective learning and discovery. These dissenters love the problem, rather than any particular solution. They want others to do the same. Like a skilled teacher, they orchestrate a discussion rather than simply trying to be the sage on the stage. Figure 7.2 summarizes effective approaches to playing the role of the devil's advocate.

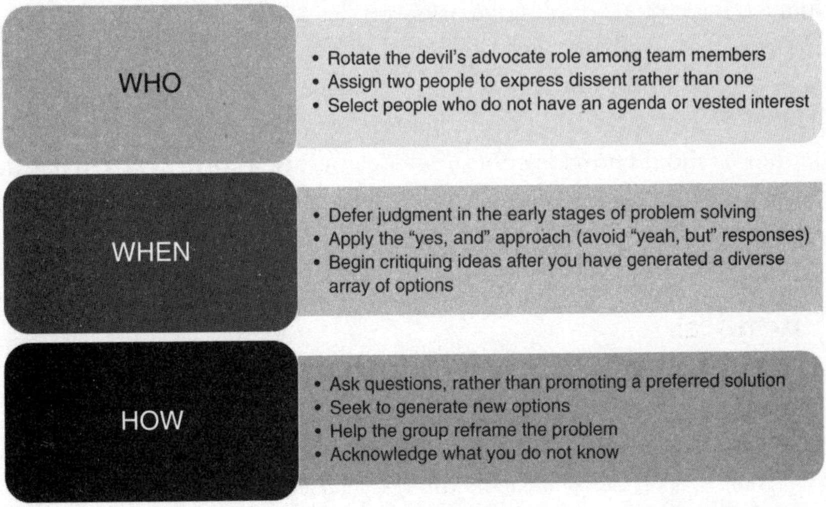

WHO	• Rotate the devil's advocate role among team members • Assign two people to express dissent rather than one • Select people who do not have an agenda or vested interest
WHEN	• Defer judgment in the early stages of problem solving • Apply the "yes, and" approach (avoid "yeah, but" responses) • Begin critiquing ideas after you have generated a diverse array of options
HOW	• Ask questions, rather than promoting a preferred solution • Seek to generate new options • Help the group reframe the problem • Acknowledge what you do not know

Figure 7.2   The Constructive Devil's Advocate

Do You Create Anything?

We often hear successful people encourage us to ignore the naysayers. Certainly, unhelpful doubters, skeptics, and critics stand in the way of creative ideas in far too many organizations. Former IBM CEO Lou Gerstner characterizes this type of climate as a "culture of no." He faced this type of unhealthy naysayer mindset when he took over at the computer maker. Gerstner described it as a "culture in which no one would say yes, but everyone could say no."[48]

In May 2010 Gawker writer Ryan Tate wrote about an incredible email exchange in which he had just taken part. He wrote: "I didn't plan to pick a fight with Steve Jobs last night. It just sort of happened: An iPad advertisement ticked me off."[49] Indeed, Tate drafted an angry email late one night, and Jobs responded much to his surprise. The two engaged in a feisty exchange for several hours. Jobs closed the discussion with a stinging rebuke, as only he could do. He wrote, "By the way, what have you done that's so great? Do you create anything, or just criticize others' work and belittle their motivations?"[50]

The best leaders encourage people to create something rather than simply playing the critic day after day. Yet, they do not eliminate dissent.

In rejecting counterproductive naysayers, leaders do not want to produce an army of sycophants. As William Wrigley, Jr. once said, "A few yes men may be born, but mostly they are made. Fear is a great breeder of them." Effective leaders encourage *helpful* critics. They think carefully about *who* should provide contrarian viewpoints, *when* dissent and debate should occur, and *how* people should play the devil's advocate in their organizations.

Endnotes

1. Ephraim Kim, *Surprise Attack: A Victim's Perspective* (Cambridge, MA: Harvard University Press, 2004), 92.

2. Bruce Riedel, "Enigma: The Anatomy of Israel's Intelligence Failure Almost 45 Years Ago," *Brookings Institution Report*, September 25, 2017 (www.brookings.edu/research/enigma-the-anatomy-of-israels-intelligence-failure-almost-45-years-ago/, accessed February 8, 2018).

3. Ibid.

4. Ibid.

5. Mitch Ginsburg, "Golda Meir: 'My heart was drawn to a preemptive strike, but I was scared'," *The Times of Israel*, September 12, 2013 (www.timesofisrael.com/golda-meir-my-heart-was-drawn-to-a-preemptive-strike-but-i-was-scared/, accessed February 8, 2018).

6. Center for Israel Education, "Agranat Commission of Inquiry Interim Report (April 1974)," 2015 (israeled.org/wp-content/uploads/2015/06/1974.4-Agranat-Commission-of-Inquiry-Interim-Report.pdf, accessed February 8, 2018).

7. Yosef Kuperwasser, "Lessons from Israel's Intelligence Reforms," *Brookings Institution Analysis Paper*, October 2007 (www.brookings.edu/wp-content/uploads/2016/06/10_intelligence_kuperwasser.pdf, accessed February 7, 2018).

8. Charlan Nemeth, Keith Brown, and John Rogers, "Devil's Advocate versus Authentic Dissent: Stimulating Quantity and Quality," *European Journal of Social Psychology*, 31(6), 2001, 707–720.

9. Personal interview with Kimball Hall, April 2010.

10. Adam Bryant, "Bob Pittman of Clear Channel on the Value of Dissent," *New York Times*, November 16, 2013 (www.nytimes.com/2013/11/17/business/bob-pittman-of-clear-channel-on-the-value-of-dissent.html, accessed February 12, 2018).

11. Adam Bryant, "Ori Hadomi of Mazor Robotics on Choosing Devils Advocates'," *New York Times,* December 24, 2011 (www.nytimes.com/ 2011/

12/25/business/ori-hadomi-of-mazor-robotics-on-choosing-devils-advocates.
html, accessed February 8, 2018).

12. Jake Knapp, John Zeratsky, and Braden Kowitz, *Sprint: How to Solve Big Problems and Test New Ideas in Just Five Days* (New York: Simon & Schuster, 2016), 35. Many of the design spring professionals that I spoke with at the Google Design Sprint Conference in November 2017 discussed the importance of the trouble-maker role. They also noted that you must choose this team member wisely.

13. Richard Banfield, C. Todd Lombardo, and Trace Wax, *Design Sprint: A Practical Guidebook for Building Great Digital Products.* (Sebastopol, CA: O'Reilly Media, 2015).

14. David Schweiger, William Sandberg, and James Ragan, "Group Approaches for Improving Strategic Decision Making," *Academy of Management Journal*, 29, 1986, 51–71; David Schweiger, William Sandberg, and Paula Rechner, "Experimental Effects of Dialectical Inquiry, Devil's Advocacy, and Consensus Approaches to Strategic Decision Making," *Academy of Management Journal*, 32, 1989, 745–772.

15. Garold Stasser has pioneered the research stream on shared information bias. For example, see Garold Stasser, and William Titus, "Pooling of Unshared Information in Group Decision-Making: Biased Information Sampling during Discussion," *Journal of Personality and Social Psychology*, 48(6), 1985, 1467–1478; Garold Stasser and Dennis Stewart, "Discovery of Hidden Profiles by Decision-Making Groups: Solving a Problem versus Making a Judgment," *Journal of Personality and Social Psychology*, 63(3), 1992, 426–434.

16. Brian Waddell, Michael Roberto, and Sukki Yoon, "Uncovering Hidden Profiles: Advocacy in Team Decision Making," *Management Decision,* 51(2), 2013, 321–340. For more information regarding my research on devil's advocacy, see Michael Roberto, *Why Great Leaders Don't Take Yes for an Answer*, 2nd edition (Upper Saddle River, NJ: FT Press, 2013).

17. Tom Kelley and Jonathan Littman, *The Ten Faces of Innovation* (New York: Doubleday, 2005), 2.

18. Theodore Roosevelt, "Citizenship in a Republic," speech delivered at the Sorbonne, Paris, France, April 23, 1910, qtd. in "The Man in the Arena," Theodore-Roosevelt.com, n.d. (www.theodore-roosevelt.com/trsorbonne speech.html, accessed February 13, 2018).

19. Steven Rattner, "A Look Inside Paul Volcker's Fed," *New York Times*, May 3, 1981 (www.nytimes.com/1981/05/03/business/a-look-inside-paul-volcker-s-fed.html, accessed February 14, 2018).

20. Inflation data drawn from: InflationData.com, "Consumer Price Index 1980–1989" and "Consumer Price Index 1970–1979," n.d. (accessed February 14, 2018).

21. William Greider, *Secrets of the Temple: How the Federal Reserve Runs the Country* (New York: Touchstone, 1987), 81.

22. Detailed documentation of all Federal Reserve Open Market Committee dissenting votes may be found at: www.stlouisfed.org/about-us/resources/a-history-of-fomc-dissents (accessed February 16, 2018).

23. Rattner, 1981.

24. Greider, *Secrets of the Temple*, 201.

25. Greider, *Secrets of the Temple*, 201.

26. Calculation performed using the Bureau of Labor Statistics inflation calculator: (data.bls.gov/cgi-bin/cpicalc.pl, accessed February 14, 2018).

27. InflationData.com, "Consumer Price Index 1980–1989."

28. Basketball-Reference.com (www.basketball-reference.com/, accessed February 15, 2018).

29. Marcel Mutoni, "Derrick Rose 'Numb' to Tom Thibodeau's Complaints," Slam, October 15, 2014 (www.slamonline.com/nba/derrick-rose-numb-tom-thibodeaus-complaints/, accessed February 15, 2018).

30. Steve Rosenbloom, "Is Thibs' Sky Falling and Are His Players Listening?" *Chicago Tribune*, October 15, 2014 (www.chicagotribune.com/sports/rosenblog/chi-tom-thibodeau-bulls-criticism-20141015-column.html, accessed February 15, 2018).

31. Adam Bryant, Kevin E. Lofton of Catholic Health Initiatives: Designate a Devil's Advocate, *New York Times*, August 8, 2015 (www.nytimes.com/ 2015/08/09/business/kevin-e-lofton-of-catholic-health-initiatives-designate-a-devils-advocate.html, accessed February 8, 2018).

32. John Zinkin, "Letter to the Editor," *Financial Times*, October 19, 2011. (www.ft.com/content/4a637670-f8e5-11e0-a5f7-00144feab49a, accessed February 15, 2018).

33. Solomon Asch, "Effects of Group Pressure upon the Modification and Distortion of Judgment," In H. Guetzkow (ed.) *Groups, Leadership and Men* (Pittsburgh, PA: Carnegie Press, 1951).

34. Ibid.

35. Serge Moscovici, Elisabeth Lage, and Martine Naffrechoux, "Influence of a Consistent Minority on the Responses of a Majority in a Color Perception Rask," *Sociometry*, 1969, 365–380.

36. Jack Arbuthnot and Marc Wayner, "Minority Influence: Effects of Size, Conversion, and Sex," *The Journal of Psychology* 111(2), 1982, 285–295.

37. Aon Hewitt, "Better Boards: Aon Research Report," 2017 (www.aon.com/unitedkingdom/attachments/retirement-investment/trustee-effectiveness/Better-Boards.pdf, accessed February 19, 2018).

38. Irving Janis, *Victims of Groupthink*, 2nd edition (Boston: Wadsworth, 1982), 141.

39. I had the opportunity to spend a day with Robert McNamara in April 2005, when he accepted an invitation from Professor Jan Rivkin and me to speak to our students at Harvard Business School. During this time, he recounted the events of the Cuban Missile Crisis in detail, including the role that Robert Kennedy and Ted Sorensen played. Throughout our discussion, he stressed the importance of constructive conflict during the decision-making process.

40. Tina Fey discussed this story during a conversation with Eric Schmidt at Google in 2011. You can watch the video here: www.youtube.com/watch?v=M8M kufm3ncc (accessed February 20, 2018).

41. Tina Fey, *Bossypants*, (New York: Reagan Arthur Books, 2011), 84.

42. I heard Jake Knapp discuss the "work alone together" philosophy at Google in November 2017. He also discusses this approach in his book. See Knapp, Zeratsky, and Kowitz, *Sprint*.

43. Scott Berkun, "In Defense of Brainstorming," ScottBerkun.com, February 13, 2012 (scottberkun.com/2012/in-defense-of-brainstorming-2/, accessed February 20, 2018).

44. William Ury. *Getting Past No: Negotiating Your Way from Confrontation to Cooperation* (New York: Bantam Books, 1993).

45. Vidya Dinamani of the LUMA Institute presented this example at a workshop that I attended at the Google Design Sprint conference in November 2017.

46. *The Paper Chase*, directed by James Bridges, Twentieth Century Fox Corporation, 1973.

47. Personal interview with Anthony Restivo, July 25, 2017.

48. Louis Gerstner, Jr., *Who Says Elephants Can't Dance?* (New York: Harper Business, 2003), 193.

49. Ryan Tate, "My Amazing Email Exchange with Steve Jobs," *Business Insider*, May 16, 2010 (www.businessinsider.com/my-amazing-email-exchange-with-steve-jobs-2010-5, accessed February 12, 2018).

50. Ibid.

CHAPTER 8

Leader as Teacher

To stimulate life, leaving it then free to develop, to unfold, herein lies the first task of the teacher.
        —Maria Montessori, Italian physician and educator

Jennifer Doudna moved to Hawaii's Big Island at the age of seven. She describes this time in her life as a "big adventure."[1] As a young girl, she treasured the opportunity to tackle a challenging problem or search for the answers to intriguing questions. Her father, a literature professor, enjoyed solving puzzles published in the local newspaper each week. For instance, he loved to decipher famous quotes printed in code. Jennifer often worked alongside him, trying to unlock the hidden messages. She says, "It really taught me about the joy of finding things out."[2]

At the age of 12, Doudna and her friend Lisa Twigg-Smith often explored the beautiful natural environment near their homes. Her classmate recalls, "We'd follow wild-pig trails, or else we'd just look at things—at mosses that were in bloom, or mushrooms."[3] Doudna became intrigued by the sleeping grass, known as hila-hila, which she discovered one day on a hike through the meadows. The leaves of this plant folded up when she touched them. Doudna says, "I'd look at that and think: Now how does that work?"[4] She became enthralled by these types of mysteries in the world around her.

Doudna's father nurtured her intellectual curiosity at every turn. He loved to learn new things, and he shared that passion with his daughter. He filled the house with books. She recalls the joy of being "the first person to know something. That, somehow, inherently was attractive to me."[5] In high school, Doudna enrolled in a chemistry class taught by Ms. Wong. She says that this inspiring teacher "taught kids about the joy of having a question about how does something work and setting up an experiment to test it."[6]

Doudna went on to earn her undergraduate degree in biochemistry at Pomona College and her doctorate at Harvard. She made her mark initially as an RNA researcher, becoming the first person to map the structure of a ribozyme. In 2011, she met a geneticist named Emmanuelle Charpentier at a conference. They began to study sequences of RNA called CRISPRs (pronounced *crispers*), clustered regularly interspaced short palindromic repeats. Doudna initially thought of this work as "a pretty small effort in my lab, just a few people having fun checking it out."[7] Intrigued as always by interesting questions, Doudna continued to experiment and learn. She did not know where this line of "curiosity-driven research" would take her. Ultimately, they achieved a major breakthrough. She thought, "Whoa, this could be transformative."[8]

Doudna and her colleagues developed a revolutionary gene-editing tool with widespread potential applications. Scientists can use this tool to discover new drugs and alter cancer cells, and someday it might help them find ways to treat genetic diseases such as cystic fibrosis. In 2015, *Time* magazine named Doudna and Charpentier two of the 100 most influential people in the world.[9] Many experts speculate that they could win a Nobel Prize someday. The journey to these extraordinary accomplishments began with a curious seven-year-old girl who learned to love a good puzzle and to ask a good question.

Not every team or organization has a potential Nobel Prize winner such as Jennifer Doudna in its midst. However, most groups contain plenty of individuals with natural curiosity and a desire to learn. Remember that the creative process often begins with a bit of wonder and a thoughtful query. Her teacher and father did not direct young Doudna to study particular topics. They nurtured her curiosity and inquisitiveness. They provided a supportive environment in which Doudna's natural skills and capabilities could flourish. As a leader, you can and should do the same for your colleagues and team members.

Building that enabling environment means transforming the mind-sets that inhibit creativity—and that is no mean feat, as you may have surmised while reading this book. That effort will take determination and persistence. Old mental models will not go away easily. As you dismantle these barriers, do not become frustrated when original ideas do not flourish immediately. Remember that the creative capabilities of people throughout your organization may have lain dormant for quite some time. Individuals often become frustrated, even cynical, over the years when their ingenuity is stifled. Their work strategy turns defensive in these circumstances: Keep my head down, do my job, don't rock the boat. You may need to light a spark, to reignite their thirst for new knowledge and their desire to develop original solutions to pressing problems.

As a leader, consider yourself a teacher at heart—not the sage on the stage imparting wisdom from on high—not that kind of teacher. Doudna's father taught literature, not science, yet he helped his daughter become a brilliant biochemist. You must cultivate and nourish curiosity and a thirst for new knowledge in your organization, much as extraordinary teachers do in their classrooms. A spirit of inquisitiveness will fuel the creative process.

Stirring the Brain

Cultivating curiosity activates the brain in profound ways. Neuroscientist Matthias Gruber and his colleagues have studied how curiosity impacts the brain and paves the way for learning to occur. In one study, they asked participants to indicate how interested they were regarding a series of questions. Later they displayed many of these questions to the research subjects once again. Individuals had 14 seconds to ponder the answer to each inquiry. Then the researchers displayed a photograph of someone's face, though it was completely unrelated to the question. Finally, the participants had the opportunity to see the answer. Twenty minutes later, the researchers conducted a surprise recognition memory test. They evaluated the extent to which individuals could remember the answers, as well as the people's faces. The scholars repeated the test a day later.

Gruber and his fellow researchers found that people performed better on this recognition memory test for those questions about which they had indicated more curiosity. That result might not surprise us. Perhaps more amazingly, though, individuals exhibited stronger recall of those disparate faces if their photos had been presented immediately after an intriguing question![10]

The researchers conducted brain imaging of the participants to learn more about how curiosity enriched learning and memory. They found enhanced activity in the region of the brain that transmitted dopamine, as well as in the hippocampus, a part of the brain critical to memory formation. Co-author Charan Ranganath explains:

> So curiosity recruits the reward system, and interactions between the reward system and the hippocampus seem to put the brain in a state in which you are more likely to learn and retain information, even if that information is not of particular interest or importance.[11]

In sum, curiosity represents a powerful form of intrinsic motivation. Teachers who activate the innate curiosity of their students tend to spur much more learning than those who simply provide stars and stickers, or other types of extrinsic rewards.

Step back then for a moment. Think about the most extraordinary teachers that instructed you as a small child. What made these women and men so special? How did they nourish your innate curiosity? As leaders, consider employing these strategies to light a spark and activate the brains of the very talented people with whom you work.

Encourage Questions

Patti Firth teaches fourth and fifth graders in Ontario, Canada. At the beginning of each unit, she brings in artifacts related to the topic—perhaps some photos and a few small objects. She hopes that these items will provoke conversation and exploration. Then Firth builds a "wonder wall" in her classroom. Or, perhaps more accurately, the students build the wall with her. She provides the young students with sticky notes and asks them to record questions that they have about the subject being examined. Firth invites them to complete the sentence, "I wonder. . ." What would they like to know about this topic? What puzzles or confuses them?[12]

As the students generate questions, they post them on the wonder wall for all to see. These questions guide the process of learning and discovery over the days and weeks to come. Make no mistake; Firth comes to the table with learning objectives of her own. However, the wonder wall invites these children to be curious and to investigate the mysteries that intrigue them. The wall demonstrates that Firth cares a great deal about the questions the children deem important, and it enables the students to share their interests and intrigue with their classmates.

University of Virginia President James Ryan explains that we all benefit from asking two types of "I wonder" questions: "I wonder why" and "I wonder if." I wonder why demonstrates intellectual curiosity. You are trying to understand what's really going on. Then, you can ask, "I wonder if. . ." In other words, what might be possible? How might we improve things? You open up your mind to new opportunities for growth, development, and change.[13]

Many teachers around the world use wonder walls these days. Peter Gamwell, a Canadian school superintendent, recalls the first time that he observed students engaged in this manner. He entered a kindergarten classroom and discovered four students seated at a table with their teacher. A nearby whiteboard displayed an array of questions about the human body, some rather silly, others very serious. Gamwell sat down with the children and asked, "What's going on here? This looks fascinating!" One child responded, "It *is* fascinating. Come join us." When the teacher pointed to the intestines on a diagram of the human body, one young boy shouted, "I know what it is! It's the intesticles!! The intesticles!" The kids erupted in laughter. The enthusiasm astounded Gamwell. He soon concluded that the concept of a wonder wall had broad applications, well beyond the kindergarten classroom.[14]

Every organization should have its version of the wonder wall. Management expert Peter Drucker would have loved this concept. He once wrote, "The most common source of mistakes in management decisions is the emphasis on finding the right answer rather than the right question."[15] What questions do your team members want to explore? The inquiries may surprise you and open up new opportunities for your organization. Moreover, your team members will be highly motivated to answer the questions that they pose, rather than always being directed to pursue problems identified from on high.

Let Them Answer

When Firth first started building wonder walls, she felt compelled to respond to each inquiry:

> When students ask questions, I really, really want to answer them. I want to share my knowledge and have them soak it all in and teach them something. I am a teacher! This is what I do!! I know stuff and teach about it!!

She learned to stop herself. She wanted them to find the answers, and to experience the joy that comes with that discovery. Firth discovered an important lesson:

> A funny thing happened. They started learning faster than I had expected. They took those questions home and found out the answers to them. They would read books during independent time and find the answers to our questions. They were discussing these things with their parents at home. It was amazing to see how excited they were about learning these concepts which in turn also allowed our discussion at school to become more vibrant and engaging.[16]

Cognitive scientist Daniel Willingham points out that the there is no fun in simply being told the answer to a challenging problem. The joy comes through the process of discovery.[17] Imagine that you have been presented a puzzle and its solution simultaneously. Do you receive any satisfaction from knowing the answer immediately? Of course not. How many times have you said, "Wait, don't tell me!" when someone tries to provide you the answer to a good riddle? You do not want them to spoil the moment. Indeed, researchers have demonstrated that people often experience a feeling of disappointment when their curiosity has been satisfied. Pleasure peaks while we anticipate a breakthrough, not after we achieve it.[18]

As leaders, presenting your answer to a problem prematurely poses several dangers, beyond simply disappointing those who love a good challenge. In many cases, leaders do not intend to impose their solution unilaterally. They want input from their team, but they discourage creative problem solving unintentionally. How does that work? Sometimes leaders will take an initial position on an issue. They present a problem to their team and describe the solution that they have in mind. Then they pose the question: "What do you think? Do you agree or disagree with this proposal?" Far too often, people stare back at them in silence, or offer subtle nods of agreement while thinking to themselves, "What a preposterous idea!"

Providing your answer before seeking input has three negative consequences. First, presenting a solution upfront may create the impression of a fait accompli. Team members may conclude that the decision has already been made, and they become irritated. They think, "Why waste our time by consulting with us? The decision is clearly pre-ordained." Second, taking an initial position means that the leader has framed the issue for the team. In so doing, he or she may constrict the range of possible solutions that the team will generate. Presenting an answer also may shape and distort the information-gathering process by triggering confirmation bias. Finally, leaders may squelch dissenting views if they present their solution before asking for others' opinions. People may not feel safe contradicting the boss's position or questioning the implicit assumptions behind that proposal.

Harry Kraemer, former CEO of Baxter International, argues that leaders need to heed the philosophy of St. Francis: Seek to understand more so than to be understood.[19] Sometimes leaders simply need to shut up and listen at the start of meetings. Ask clarifying questions, but do not interrupt to articulate your viewpoint. Quietly take detailed notes, and later play back what you heard to confirm that you have comprehended others correctly. Vicki Williams, an executive at NBCUniversal, reminds leaders throughout her organization that, "Leaders who don't listen will soon find themselves with people who have nothing to say."[20]

Canadian entrepreneur Cameron Herold advocates speaking last in team meetings if you are the leader. In addition, he recommends calling on junior members of the team before requesting input from senior executives. He explains:

> When you have quieter, more reserved people in a meeting, the best thing you can do as the leader is hold your ideas back until the end. Too often, leaders offer their ideas first. But people don't become confident, or grow as leaders, by listening to what you have to say. Instead, you need to encourage members of the team to offer their ideas first, especially those less inclined to speak up.[21]

Share Failure Stories

Children sometimes do not pursue a particular field of study because they lack confidence. They suppress their innate curiosity because they believe that achievement in a particular domain requires extraordinary intellectual ability that they do not possess. Listening to heroic stories of incredible accomplishment only exacerbates the problem.

Ron Gray, a faculty member at the University of Northern Arizona, used to teach science to middle-school students. He believes in telling young people the hard truths about the messy process of scientific discovery. He wants them to know about the obstacles and dead-ends encountered by famous people such as Albert Einstein. Gray hopes that students will realize that these remarkable men and women were human and made mistakes just as we all do. That recognition, he believes, will enable young people to see themselves as capable of doing great scientific work.[22]

Recent research by Xiaodong Lin-Siegler and her colleagues demonstrates that children learn more effectively when they appreciate the hardships and struggles of people who ultimately attained great discoveries. They asked more than 400 high-school students to read stories about famous scientists. Some students read stories about the mistakes that Albert Einstein, Marie Curie, and Michael Faraday made at one point or another. The narratives recounted how these scientists learned and recovered from their failures on the path to a remarkable breakthrough. Other students read stories about the personal hardships experienced by these three scientists. For example, they learned about how Faraday came from a very poor family and encountered religious prejudice throughout his life. The third group of students read heroic tales of achievement by these three individuals, with little discussion of failures or adversity.

The researchers tracked the students' grades in their science class before and after they read these stories. The average academic performance of the three groups did not differ prior to the study. Remarkably, the students who read stories about intellectual struggles and personal adversity exhibited higher science grades in the subsequent six-week period than the students who read heroic stories of scientific genius. Most importantly, the lowest-performing students experienced the biggest boost in performance after learning about the messiness of the scientific process.[23]

Leaders should think about sharing their own stories of adversity and failures with colleagues and subordinates. Organizational learning expert Amy Edmondson argues that the most effective leaders model curiosity for team members and acknowledge their own fallibility. In so doing, they make it safe for team members to experiment and make mistakes. Individuals become comfortable pursuing the issues and questions that arouse their curiosity, because they do not fear being blamed or punished if their experiment fails.[24]

Sara Blakely, founder and CEO of Spanx, remembers conversations with her father as a child. He often would ask about her recent failures. Blakely recalls telling him, "Dad, Dad, I tried out for this, and I was horrible!" He did not express disappointment in his daughter. Blakely loved his response, "He would actually high-five me and say, 'Congratulations, way to go!' Failure for me became not trying, versus the outcome."[25]

Blakely has tried to send the same message to everyone in her organization. Several years ago, she played the Britney Spears song, "Oops, I Did It Again" at a company-wide meeting. Then she recounted a series of errors that she had made as Spanx grew and prospered. Blakely sent a clear message to her team: I'm fallible, too. I understand that failure will occur when people try new things.[26]

Celebrate Mistakes

When Maxine Clark was in third grade, she received a startling comment on her report card. The teacher wrote that she asked too many questions in class. Clark admits, "I was a really nosy kid. I always wanted to understand why people did this, or why people did that."[27] After reading the report card, Clark's mother became irate, not with her daughter, but with the teacher. She marched down to the elementary school. It is the only time Clark can ever remember her mom complaining about a teacher's conduct. Her mother encouraged the teacher to talk to a colleague at the school, Mrs. Grace, who had instructed Clark in the first grade. Mrs. Grace had a very different philosophy about curiosity and learning.

In Clark's first-grade classroom, Mrs. Grace offered a red pencil award each week. Now, most of us do not have fond memories of red pencils. After all, our teachers often returned our graded homework with our errors marked clearly in red. Mrs. Grace, though, chose to provide her award to the student who had raised his or her hand often in class that week.[28] Clark recalls:

> Mrs. Grace always had a way of making you special. When you made the best mistake of the week in her class, she gave you her red pencil. That was a coveted prize because when you're a first grader, you have those thicker pencils, and she had this nice, thin, really sharp red pencil.[29]

Mrs. Grace encouraged her students to ask questions and not to fear making mistakes. She wanted them to take risks in her class, in pursuit of new knowledge.

Clark learned a great deal from many teachers in her life, but she remembers Mrs. Grace quite fondly even today. Her curiosity, nurtured in that first-grade classroom, served her well throughout her career. Roughly four decades later, Clark came up with a novel idea for enabling children to create their own toys. She founded Build-A-Bear Workshop, and the retailer grew rapidly and became very successful. As chief executive, Clark decided to offer the Red Pencil Award to her employees at Build-A-Bear, celebrating those courageous individuals who questioned existing ways of working, experimented with new practices, and learned from their mistakes.

Empathize Genuinely

David Glahe teaches second grade at Hyde Park Elementary School in Niagara Falls, New York. Many families in this school district have low incomes. In fact, most children qualify for free lunches. For more than 20 years, Glahe has written a letter each summer to the parents of his incoming students. He introduces himself and offers to meet with them before the start of the school year. The meetings are purely voluntary. Unlike most teachers, he does not ask the parents to come to his class-room. Instead, he offers to visit the families in their homes, and many parents accept his unconventional offer.

Glahe uses the home visits to get to know the children, and he comes prepared with a set of questions to start the discussion. He notes, how-ever, that the most important question is: "What else should I know about your child?" Parents tell him about their children's interests and passions, as well as their struggles. Glahe says, "They are normally very forthcoming. What they tell me runs the gamut. When they tell me about the child's interests, that's really valuable because it helps me make connections that can help with learning."[30] Through these visits, Glahe gains an appreciation for the challenging circumstances many of his stu-dents face in their homes and neighborhoods. Principal Mary Kerins notes, "He makes connections with the parents, and builds relationships with his students. They know that he cares about them."[31]

Glahe's actions call to mind the words of Theodore Roosevelt, who once said, "Nobody cares how much you know until they know how much you care." We all remember inspiring teachers who went to extraordinary lengths to take an interest in our development and well-being. We did not work hard in their class because they held degrees from elite educational institutions or had received major awards for their academic accomplishments. We put forth extra effort because they cared.

Efforts to empathize with others not only help us understand the challenges that they face and demonstrate that we care about them. These outreach efforts help us discover things we have in common that we otherwise may not have uncovered. Consider a 2015 study by Hunter Gehlbach and his colleagues. They created a "get to know you" sur-vey for 315 high school students and their teachers. Afterward, some students and teachers received feedback regarding five things that they had in common with one another. The teachers felt they had devel-oped better relationships with their students after learning about shared interests and experiences. These teachers and students interacted more often after learning about their similar interests. Preliminary evidence from this study even suggests that academic achievement increases when teachers and students discover what they have in common with one another. The minority achievement gap may decrease as well in these circumstances.[32]

Leaders across all types of organizations should consider how they can empathize genuinely with their employees. Great new ideas may emerge from these efforts. Chris Nassetta took over as CEO of Hilton in 2007. He had to engineer a turnaround at the struggling hotel chain. Nassetta decided that his executives needed to connect more closely with the associates on the front lines to understand their concerns, identify their frustrations, and hear their ideas. He worried that the top management team had lost touch with those doing the real work. Nassetta remembered learning a great deal in his first job in the hotel business at age 18, cleaning toilets at the Capitol Holiday Inn in Washington, D.C. He decided that it was time to return to his roots.

Nassetta directed his top managers to spend one week per year working at one of the Hilton hotels around the world. He did the same, thereby modeling the behavior he expected from his team members. They took on housekeeping tasks, helped the facilities and grounds crew perform maintenance, and greeted guests at the front desk.[33] Nassetta notes, "Their job is harder than your job. You get in there, and you pay them the respect."[34] The management team learns a great deal from this "immersion" process each year, and new ideas emerge from the many conversations that take place between executives and frontline workers.

Make Them Believe

Daniel Willingham has asked many people "Who was the most important teacher in your life?" He notices a common theme among their responses:

> First, most people have a ready answer. Second, the reason that one teacher made a strong impression is almost always emotional. The reasons are never things like, "She taught me a lot of math." People say things like, "She made me believe in myself" or "She taught me to love knowledge."[35]

Extraordinary teachers set high expectations, and they demand a great deal from their students. They challenge them deeply. However, they demonstrate a great deal of confidence in these children. They express a sincere belief that students can elevate their performance and achieve things they never imagined. These outstanding teachers do not praise children at every turn. Instead, they take the time to provide feedback, illustrate how children can improve, and call out students when they put forth less-than-stellar effort.

David Yeager and his colleagues conducted a fascinating intervention to demonstrate how a teacher's belief in his or students can impact performance. They worked with a group of middle-school teachers and students in three social studies classrooms. The teachers asked the students to write essays. Afterward, the teachers provided detailed feedback, as they would normally for their students. The researchers intervened, though, to add one line to the feedback that the teachers generated. Some students received a note at the bottom of their papers saying, "I'm giving you these comments because I have very high expectations and I know that you can reach them." Others simply received a note explaining, "I'm giving you these comments so that you'll have feedback on your paper." Students in this study had an opportunity to revise their essays after receiving feedback, though the teachers did not require them to do so.

The students who received the note about high expectations chose to revise their essays at a significantly higher rate than those who received the generic note. Moreover, these same students received higher evaluations on their essays from their teachers as well as independent coders. The researchers discovered a fascinating effect with regard to the minority achievement gap. They report that African-American students benefited from the high expectations feedback more than white students, with the greatest impact of all among African-American children who felt a sense of mistrust about school.[36]

Yeager's work connects back to our previous discussion about empathy. He argues that teachers have to try to see the learning process through their students' eyes. He explains that we must understand how and why students' concerns, worries, and assumptions impede the learning process. Of course, a similar statement can be made about employees in any organization. Leaders must step back and consider the beliefs and worries that prevent people from asking questions, proposing new ideas, and trying new things.

Introduce Novelty

As a child, when I entered Mrs. Marois's classroom, I never knew quite what to expect. One day, we might be studying Mary Cassatt and Claude Monet. Later that week, I might be exploring the invention and history of the automobile. Some classes involved excursions into the woods behind our school to explore nature. From time to time, she asked us to lie down on the carpet and close our eyes. As she read stories, Mrs. Marois asked us to imagine ourselves in some faraway land. She even gave me the freedom to complete a project about the history of professional football, much to the chagrin of my mother. We did not just learn about a variety of topics in Mrs. Marois' classroom; we experienced a diverse array of teaching and learning methods. The surprises that awaited us each day always created a sense of excitement. What would Mrs. Marois have up her sleeve today?[37]

Exceptional teachers typically do not employ the same pedagogical approaches day after day. They use a variety of learning techniques. Their classrooms often come alive when they surprise students with a new activity. They also expose their students to an array of new experiences. They take them to places they have never been, introduce them to people they have never met, and expose them to ideas that challenge their entrenched ways of thinking. Sometimes, these activities make the students uncomfortable. However, the exposure to new people, situations, and ideas often stimulates their curiosity and motivates them to gain more knowledge.

Scientists Nico Bunzeck and Emrah Düzel have studied how the brain responds to novel situations. In particular, they have focused on the substantia nigra/ventral tegmental area of the brain. Using brain imaging, their research shows that this region responds more vigorously to novel stimuli than familiar ones. They have concluded that novelty enhances exploration and learning.[38]

How can leaders introduce novelty as a means of stimulating curiosity, inquiry, and knowledge creation in their organizations? Provide new experiences for your employees. Ask them to travel to new regions or learn about different cultures. Encourage them to read broadly, not simply within their technical domain. Offer opportunities for employees to take classes in fields not directly related to their current position. Provide challenging assignments outside of the functional area in which they have specialized to that point.

Novelty will pay dividends, though perhaps not immediately. Remember, though, that many creative breakthroughs occur when individuals make connections between seemingly disparate concepts. Those links and relationships may not become apparent overnight. Sometimes, it seems as though these breakthroughs are simply the product of luck. On the contrary, Harvard scholar Ethan Zuckerman argues that, "Engineering serendipity is this idea that we can help people come across unexpected but helpful connections at a better than random rate."[39] How do we do that? We have to encourage our team members to become purposeful about seeking novel learning experiences, rather than only deepening their knowledge in their particular field of specialization.

When Tor Myhren served as a senior executive at advertising giant WPP, he once banned meetings from nine o'clock to noon on Thursday mornings. One of his direct reports, Gina Sclafani, describes this period each week as "officially sanctioned time for us to expand our minds and devote time to ideas that kept getting pushed to the side, using whatever methods worked for us."[40] Sclafani decided to use some of this time to learn something new, particularly to investigate a topic about which she did not have a great deal of interest. She mistakenly thought that she would enjoy this time to explore and expand her mind. Sclafani concluded, "I was wrong. Going outside your comfort zone is—and this should have been obvious–uncomfortable. Even painful."[41] However, she made connections among seemingly unrelated topics, adopted new perspectives on issues, and came to question some of her assumptions. Sclafani soon found herself approaching problems differently as a creative director in the advertising business. Novelty proved disconcerting

Table 8.1　What Great Teachers and Leaders Do to Ignite Curiosity

Method	Description
Encourage Questions	Build your version of the "wonder wall." What questions do your colleagues want to investigate? What challenges excite them?
Let Them Answer	Resist offering your solutions upfront. Give others an opportunity to voice their ideas before presenting your views.
Share Failure Stories	Speak openly about how messy and challenging the creative process can be, including the failures that you have experienced.
Celebrate Mistakes	Reward and honor those who dare to try and to experiment, and who learn from their mistakes and failures.
Empathize Genuinely	Learn about the challenges and obstacles that your colleagues face. Try to walk in their shoes, even if just for a short period of time.
Make Them Believe	Set high expectations and demand great work, but demonstrate that you believe in their talents and capabilities.
Introduce Novelty	Stimulate the brain by providing novel experiences, exposing others to new ideas, and introducing them to people with different backgrounds and perspectives.

at times, but it stimulated her brain. She concluded that the pain was worth it. Review Table 8.1 for a summary of how leaders can stimulate curiosity.

The Leader's Greatest Reward

Do you know how to make a teacher's day? Send them a brief note expressing your appreciation. Tell them how and why they made an impact on you. Last year, a former student dropped me an email describing an amazing new career opportunity that she had earned. She wrote, "I don't thank you enough, but I don't think I would ever be on this path if it weren't for you." My eyes teared up, and I went through the rest of the day with a spring in my step. Without a doubt, teachers attain their greatest rewards through witnessing the accomplishments of their students, both in the classroom and beyond.

Leaders should think about their accomplishments in the same way. Yes, we all find it exhilarating when we solve a tough problem ourselves. What happens, though, when we enable others to achieve a creative breakthrough? Think about how rewarding and gratifying you will feel as a leader when ideas flourish throughout your organization. Enabling others to explore, experiment, learn, and create is your duty as a leader, and it's potentially the most rewarding work you will ever do.

Your team members and colleagues have a tremendous capacity for creativity. Never forget that. As leaders, you must marshal the collective intellect of your people. Like our most beloved teachers, the best leaders build supportive environments that enable people to prosper. With your help, individuals can employ their creativity to solve the most "wicked" problems any organization could face. You do not have to have all the answers. You must nourish the innate curiosity of your colleagues. You need to dismantle the barriers and transform the mindsets that impede others' creativity. If we clear the path, people will race in new directions and discover new destinations. What new ideas will flourish or new opportunities will arise if you simply take down the obstacles ahead of them? I wonder . . .

Endnotes

1. Rajendrani Mukhopadhyay, "'On the Same Wavelength'," ASBMB Today, August 2014 (www.asbmb.org/asbmbtoday/201408/Features/Doudna/, accessed February 26, 2018).

2. Melissa Pandika, "CRISPR Pioneer Jennifer Doudna Shares Her Outlook for the Groundbreaking Gene-Editing Tool," *Chemical & Engineering News* 95(17), April 18, 2017, 28–29 (cen.acs.org/articles/95/i17/CRISPR-pioneer-Jennifer-Doudna-shares-her-outlook-for-the-groundbreaking-gene-editing-tool.html, accessed February 26, 2018).

3. Jennifer Kahn, "The Crispr Quandary," *The New York Times Magazine*, November 9, 2015 (www.nytimes.com/2015/11/15/magazine/the-crispr-quandary.html, accessed February 26, 2018).

4. Ibid.

5. Mukhopadhyay, "'On the Same Wavelength.'"

6. Ibid.

7. Ibid.

8. Kahn, "The Crispr Quandary."

9. "The 100 Most Influential People 2015," *Time*, April 16, 2015 (time.com/collection/2015-time-100/, accessed February 26, 2018).

10. Matthias Gruber, Bernard Gelman, and Charan Ranganath. "States of Curiosity Modulate Hippocampus-Dependent Learning via the Dopaminergic Circuit," *Neuron*, 84(2), 2014, 486–496.

11. Andy Fell, "Curiosity Helps Learning and Memory," Egghead blog, UC Davis, October 2, 2014 (blogs.ucdavis.edu/egghead/2014/10/02/curiosity-helps-learning-and-memory/, accessed February 27, 2018).

12. Patti Firth, "Talking Inquiry: Making a Wonder Wall," MadlyLearning.com, August 6, 2015 (www.madlylearning.com/wonderwall/, accessed March 1, 2018).

13. James Ryan, *Wait, What? And Life's Other Essential Questions* (New York: HarperOne, 2017).

14. Peter Gamwell and Jane Daly, *The Wonder Wall: Leading Creative Schools and Organizations in an Age of Complexity* (Thousand Oaks, CA: Corwin Press, 2017), 2.

15. Peter Drucker, *The Practice of Management* (New York: Harper Row, 1954), 351.

16. Firth, "Talking Inquiry: Making a Wonder Wall."

17. Daniel Willingham, *Why Don't Students Like School* (New York: John Wiley & Sons, 2009).

18. George Lowenstein, "The Psychology of Curiosity: A Review and Reinterpretation," *Psychological Bulletin,* 116(1), 1994, 75–98.

19. I heard Harry Kraemer share this anecdote at a lecture he presented in September 2017.

20. Jennifer V. Miller, "Leaders Who Listen Create Space for Great Ideas to Emerge," SmartBrief.com, June 13, 2017 (www.smartbrief.com/original/2017/06/leaders-who-listen-create-space-great-ideas-emerge, accessed March 2, 2018).

21. Cameron Herold, "Why Leaders Should Speak Last in Meetings," *The Globe and Mail*, March 24, 2017 (www.theglobeandmail.com/report-on-business/careers/leadership-lab/why-leaders-should-speak-last-in-meetings/article31934105/, accessed March 2, 2018).

22. Marjee Chmiel, "Stories of Challenge and Triumph: Using History of Science in Our Science Teacher," Smithsonian Science Education Center website, n.d. (ssec.si.edu/stemvisions-blog/stories-challenge-and-triumph-using-history-science-our-science-teacher, accessed March 3, 2018).

23. Xiaodong Lin-Siegler, Janet Ahn, Jondou Chen, Fu-Fen Anny Fang, and Myra Luna-Lucero, "Even Einstein Struggled: Effects of Learning about Great Scientists' Struggles on High School Students' Motivation to Learn Science," *Journal of Educational Psychology* 108(3), 2016, 314–328.

24. Amy Edmondson, *Teaming: How Organizations Learn, Innovate, and Compete in the Knowledge Economy* (San Francisco: Jossey-Bass, 2014).

25. Kathleen Elkins, "The Surprising Dinner Table Question That Got Billionaire Sara Blakely to Where She Is Today," *Business Insider*, April 3, 2015 (www .businessinsider.com/the-blakely-family-dinner-table-question-2015-3, accessed March 5, 2018).

26. Shana Lebowitz, "A Self-Made Billionaire Explains How Britney Spears Helped Her Teach a Key Business Lesson to Her Employees," *Business Insider*, June 22, 2016 (www.businessinsider.com.au/sara-blakely-teaches-spanx-employees-to-embrace-failure-2016-6, accessed March 5, 2018).

27. Tom Peters Company, "Clark, Maxine," Tom Peters Company website, 2006 (tompeters.com/cool-friends/clark-maxine/, Accessed March 6, 2018).

28. Maxine Clark and Amy Joyner, *The Bear Necessities of Business: Building a Company with Heart* (New York: Wiley, 2007).

29. Tom Peters Company, "Clark, Maxine."

30. Anne Neville, "Award-Winning Niagara Falls Teacher Learns from Parents During House Calls," *The Buffalo News*, February 1, 2018 (buffalonews.com/2018/01/31/award-winning-teacher-learns-a-lessons-from-parents/, accessed March 6, 2018).

31. Ibid.

32. Hunter Gehlbach, Maureen Brinkworth, Aaron King, Laura Hsu, Joseph McIntyre, and Todd Rogers, "Creating Birds of Similar Feathers: Leveraging Similarity to Improve Teacher–Student Relationships and Academic Achievement," *Journal of Educational Psychology*, 108(3) 2016, 342–352.

33. I first learned about this initiative during a conversation with Kimo Kippen, former chief learning officer of Hilton hotels, at a meeting of the Human Resources Leadership Forum in Arlington, Virginia in December 2012.

34. Scott Mayerowitz, "How Hilton's CEO Led the Company's Massive Turnaround," *Inc.*, July 30, 2014 (www.inc.com/associated-press/how-hilton-ceo-turned-around-his-hotel-business.html, accessed March 3, 2018).

35. Willingham, *Why Don't Students Like School*, 144.

36. David Yeager, Valerie Purdie-Vaughns, Julio Garcia, Nancy Apfel, Patti Brzustoski, Allison Master, William Hessert, Matthew Williams, and Geoffrey Cohen, "Breaking the Cycle of Mistrust: Wise Interventions to Provide Critical Feedback across the Racial Divide," *Journal of Experimental Psychology: General*, 143(2), 2014, 804–824.

37. Kathy Marois taught at Essex Elementary School for decades. I am grateful for the tremendous impact that she had on me and many other children.

38. Nico Bunzeck and Emrah Düzel, "Absolute Coding of Stimulus Novelty in the Human Substantia Nigra/VTA," *Neuron*, 51(3), 2006, 369–379.

39. David Scharfenberg, "MIT's Zuckerman on Building a More Cosmopolitan Internet," WBUR News, July 17, 2013 (legacy.wbur.org/2013/07/17/zuckerman-rewire-interview, accessed March 8, 2018).

40. Gina Sclafani, "The Creative Benefits of Exploring the Uncomfortable," *Fast Company*, May 10, 2012 (www.fastcompany.com/1680766/the-creative-benefits-of-exploring-the-uncomfortable, accessed March 12, 2018).

41. Ibid.

RESOURCES

This list of instructional materials may be useful for teaching about how leaders can stimulate creativity, either in the classroom or in corporate leadership development programs.

Case Studies

Ryan Buell and Andrew Otazo, "Human-Centered Service Design," Harvard Business School Case Study 9–615–022, January 29, 2016.

Roy Chua and Robert Eccles, "Managing Creativity at Shanghai Tang," Harvard Business School Case Study 9–410–018, August 3, 2009.

M. Julia Prats, Javier Quintanilla, and Jordan Mitchell, "elBulli's Magic Recipe," IESE Business School Case Study 0–608–017, October 13, 2008.

Stefan Thomke and Barbara Feinberg, "Design Thinking and Innovation at Apple," Harvard Business School Case Study 9–609–066, May 1, 2012.

Simulations

Michael A. Roberto, *New Venture Exercise: The Food Truck Challenge*, Harvard Business Publishing, 2016.

Michael A. Roberto and Amy C. Edmondson, *Everest Leadership and Team Simulation – Version 3.0*, Harvard Business School Publishing, 2017.

Toolkits and Handbooks

IDEO's Human-Centered Design Toolkit (www.ideo.com/post/design-kit).

Stanford d.school's Design Thinking Bootleg (dschool.stanford.edu/resources/the-bootcamp-bootleg).

Google's Design Sprint Kit (designsprintkit.withgoogle.com/).

Luma Institute's Innovating For People Handbook Of Human-Centered Design Methods (www.luma-institute.com/products-and-services/workshops/fundamentals-innovation-human-centered-design/).

Articles

Allison G. Butler and Michael A. Roberto, "When Cognition Interferes with Innovation: Overcoming Cognitive Obstacles to Design Thinking." *Research-Technology Management*. 61, no. 3 (2018): 45–51.

Ed Catmull, "How Pixar Fosters Collective Creativity," *Harvard Business Review*, 86, no. 9 (2008): 64–72.

Linda Hill, Greg Brandeau, Emily Truelove, and Kent Lineback, "Collective genius," *Harvard Business Review* 92, no. 6 (2014): 94–102.

Tom Kelley and David Kelley, "Reclaim your creative confidence," *Harvard Business Review* 90, no. 12 (2012): 115–118.

Jennifer Riel and Roger Martin, "An integrative methodology for creatively exploring decision choices," *Strategy & Leadership* 45, no. 5 (2017): 3–9.

TED Talks

Tim Brown, "Tales of Creativity and Play," 2008.

Julie Burstein, "4 Lessons in Creativity," 2012.

Elizabeth Gilbert, "Your Elusive Creative Genius," 2009.

Steven Johnson, "Where Good Ideas Come From," 2010.

Ken Robinson, "Do Schools Kill Creativity?" 2006.

Amy Tan, "Where Does Creativity Hide?" 2008.

ACKNOWLEDGMENTS

This journey to learn more about how to stimulate creativity in organizations began in 2012. My colleagues and I created the Bryant University IDEA program, a unique immersive experience in which every first-year student learns design thinking in a hands-on, experiential manner. I owe a huge debt of gratitude to the colleagues who have shaped and influenced my thinking about creativity through that project: Allison Butler, Lori Coakley, Rich Holtzman, Rich Hurley, Matt Kreimeier, Amanda McGrath, Chris Morse, Kristin Ridge, Jim Segovis, and Sue Zarnowski. Thank you to all the students who served as IDEA mentors and leadership team members over the years. I learned a great deal from each of you and enjoyed working together on this amazing program. Allison Butler deserves special recognition; our work together has had a profound influence on my thinking about creativity. Lori Coakley and Jim Segovis merit special mention too, as I have learned so much by partnering with these two incredibly talented teacher-scholars.

Many people have helped me learn about creativity through the lens of design thinking. Thank you to Kai Haley for inviting me to Google's first Design Sprint Conference and enabling me to learn from talented creative professionals at so many different organizations. Thank you to the team at IDEO for hosting my visit so that I could learn more about their unique approach to human-centered innovation, and for sharing their methodologies openly with educators around the world. Many others spent time with me discussing their approach to design thinking including Andrew Webster at Experience Point, Phil Gilbert at IBM, Laura Ramos at Gannett, Michelle Proctor at FedEx, Saul Kaplan at

Business Innovation Factory, and numerous other experts across a range of industries.

I am indebted to Ed Catmull for spending a few hours with me one afternoon in New York City, so that I could learn about creativity at Disney and Pixar. Thank you to Brian Terkelsen for speaking me with about his creative collaboration with Mark Burnett in the early days of reality television, and to Ed Viesturs and David Breashears for many conversations about leading Everest expeditions over the years. I also learned from conversations with Anthony Restivo at DraftKings, Jason Park at Allstate Digital Ventures, Bill Pacheco at Keurig, Bill Ribaudo at Deloitte, Major Mike Manning of the Rhode Island National Guard, Jane Souza and Nik Patel at Fidelity Investments, Jeff Gagnon at Amica Insurance, the teachers and students at Montrose School and Lincoln School, and so many others.

Thank you to the colleagues who collaborated with me to create the Food Truck Challenge, an exercise designed to teach about iteration and prototyping. They include Michael Bean and the team at Forio, as well as Nicole Harris, Navid Sharifi, Lin Mahoney, Mark Fuller, and Abby Vargas at Harvard Business Publishing. Our collaboration has been a highly rewarding learning-by-doing experience. Amy Edmondson warrants special recognition for teaching me so much about psychological safety and effective teams through our research, writing, and teaching efforts together. Brian Waddell and Sukki Yoon helped advance my thinking about devil's advocacy through our collaboration on an honors thesis and journal article published several years ago.

Glenn Sulmasy, Provost at Bryant University, and Madan Annavarjula, Dean of the College of Business, provided me the opportunity to take a full-year sabbatical to work on this book project. This past year has been an enriching and rewarding intellectual experience. I appreciate their support. Thank you to Jeanenne Ray, Vicki Adang, Christina Verigan, Sharmila Srinivasan and the entire team at Wiley for taking enthusiastic interest in my work and helping me to publish this book. I am fortunate to have had the opportunity to collaborate with them.

Most importantly, I would like to express my appreciation to my family. Our three children, Grace, Celia, and Luke, have taught me more about creativity and curiosity than any research project possibly could. I am so proud to be your dad. Thank you to my parents for being bold risk-takers who came to America in search of opportunity for their children. Finally, my heart belongs forever to an incredible woman who inspires creativity in her students every day and shares my passion for the vocation of teaching. Thank you, Kristin, for your unconditional love and support.

ABOUT THE AUTHOR

Michael A. Roberto is the Trustee Professor of Management at Bryant University in Smithfield, Rhode Island. He previously

185

served on the faculty at Harvard Business School and as a visiting professor at NYU's Stern School of Business.

Professor Roberto ranked #25 on the Case Centre's most recent list of the best-selling case study authors in the world. He also has created simulations adopted by business schools and companies around the world including the *Everest Leadership and Team Simulation* and the *Food Truck Challenge*, both in collaboration with Forio and Harvard Business Publishing.

He has written two previous books: *Why Great Leaders Don't Take Yes For An Answer* (2nd edition published in 2013), and *Know What You Don't Know* (2009). He also has developed several audio/video lecture series for The Great Courses on the topics of decision making, leadership, and business strategy.

Professor Roberto received an A.B. in economics from Harvard College in 1991. He earned an M.B.A. from Harvard Business School in 1995, graduating as a George F. Baker Scholar. He also received his doctorate from Harvard Business School in 2000.

You can learn more about Professor Roberto at his website (www .professormichaelroberto.com), and you can read his blog at http:// michael-roberto.blogspot.com/.

INDEX

Page references followed by f indicate an illustrated figure; followed by t indicate a table.